T0181913

High Voltage Digital Power Line Carrier Channels

Anton G. Merkulov • Yuri P. Shkarin
Sergey E. Romanov • Vasiliy A. Kharlamov
Yuri V. Nazarov

High Voltage Digital Power Line Carrier Channels

Anton G. Merkulov
Almaty University of Power Engineering
and Telecommunications
Almaty, Kazakhstan

Yuri P. Shkarin
Power System and Network Research and
Design Institute "Energosetproject"
Moscow, Russia

Sergey E. Romanov
Unitel Engineering LLC
Moscow, Russia

Vasiliy A. Kharlamov
Unitel Engineering LLC
Moscow, Russia

Yuri V. Nazarov
NPF Modem LLC
Saint Petersburg, Russia

ISBN 978-3-030-58367-5 ISBN 978-3-030-58365-1 (eBook)
https://doi.org/10.1007/978-3-030-58365-1

This Springer imprint is published by the registered company Springer Nature Switzerland AG
The registered company address is: Gewerbestrasse 11, 6330 Cham, Switzerland

Preface

Power line carrier (PLC) communication along high voltage (HV) and extra high voltage (EHV) electric power transmission lines (PL) is a common type of telecommunication used in the electric power industry. During more than 100 years of PLC, theoretical basis for designing PLC channels, methods for determining high frequency (HF) path parameters and noise characteristics, coupling schemes, principles for constructing HF paths over phase conductors, conducting ground wires, and intra-phase and intra-cable connections have been developed. PLC technology has developed to meet current industry requirements: principles of PLC equipment designing, modulation methods, and the theory of signal processing and coupling devices have changed significantly. This book considers PLC channels along 35–750 kV alternating current (AC) electric power transmission lines.

While digital fiber-optic and radio communication systems are widely used, attitude of specialists to using digital PLC systems (DPLC) in the power industry is ambiguous. Some of them say that the technology is outdated and unsuitable for modern electric grids due to the low bandwidth of channels, their exposure to weather factors, and commutation interference. Others state that PLC will remain a reliable telecommunication mean in the electric power industry for many years due to full coverage of the country's territory with power transmission lines, the highest reliability of the transmission medium (phase conductors of power transmission lines), and relatively low capital costs for construction and operation of long PLC channels.

A serious problem in operating and design organizations is the lack of specialists working in DPLC systems. Use of modern equipment requires specialists to have appropriate high-quality theoretical knowledge in several areas. First, it is necessary to know specific methods for creating PLC paths and measuring their parameters. No matter how perfect the PLC equipment is, if there are problems in the line, it is impossible to build a reliable communication channel. Secondly, it requires knowledge in telecommunications, understanding of principles of signals multiplexing, operation of modems, multiplexers, subscriber tone terminal units, and various data transfer interfaces. Due to the fact that modern PLC systems are often used to transmit packet traffic of IP networks, knowledge in information technology (IT) is necessary. It should be noted that the problems associated with DPLC channels are not described enough in available sources. The purpose of this book is to fill this gap.

The book consists of five chapters which include information about designing digital PLC terminal equipment (Chap. 1), modems (Chap. 2), multiplexers and network elements of DPLC systems (Chap. 3), influence of HF path parameters on DPLC channel operation (Chap. 4), and design of DPLC channels (Chap. 5). The content of the book is based on the extensive practical experience of authors in the area of PLC channels creation, HF path calculations, development, and implementation of new approaches to designing DPLC for electric grids.

The book is intended for researchers and specialists of design and operating organizations engaged in practical issues of creation and operation of communication networks based on digital PLC channels. It will be useful for teachers, postgraduates, and students of communications and power engineering.

Almaty, Kazakhstan — Anton G. Merkulov
Moscow, Russia — Yuri P. Shkarin
Moscow, Russia — Sergey E. Romanov
Moscow, Russia — Vasiliy A. Kharlamov
Saint Petersburg, Russia — Yuri V. Nazarov

Abbreviations

ACK	acknowledgment
APLC	analog power line carrier
AWGN	additive white Gaussian noise
BER	bit error rate
CC	coupling capacitor
CD	coupling devices
CPL	cable power line
COPL	cable overhead power line
DMT	discrete multi-tone
DPLC	digital power line carrier
DTS	digital transmission system
HC	header compression
HF	high frequency
IHRD	ice hoarfrost and rime deposits
IP	internet protocol
LAN	local area network
LMU	line matching unit
LT	line trap
MOS	mean opinion score
MSS	maximum segment size
MTU	maximum transport unit
MUX	multiplexor
OFDM	orthogonal frequency division multiplexing
OPL	overhead power line
PAPR	peak to average power ratio
PEP	peak envelope power
PLC	power line carrier
PPP	point to point
PSK	phase shift keying
QAM	quadrature amplitude modulation
RFC	request for comments

ROHC	robust header compression
RTP	real time protocol
RTT	round trip time
SNMP	simple network management protocol
TCM	trellis-coded modulation
TCP	transport control protocol
UDP	user datagram protocol
VoIP	voice over IP
WAN	wide area network
WB-LSB	window based least significant bit encoding
A	amplitude
a	attenuation
B	bit rate, transmission capacity
C	capacitance
D	diameter
E	energy
f	frequency
I	current
L, l	length
M	modulation index
P	power
p	power level
$P(X)$	probability
R	resistance
Rinf	line rate
S	size of field in pocket
t	time
U	voltage
υ	propagation velocity
W	bandwidth
Z	impedance
α	attenuation constant
β	phase constant
γ	propagation constant
η	spectral efficiency
δ	square complex n-th order matrix for transformation of a system of mode current vectors into a system of phase current vectors
λ	square complex n-th order matrix for transformation of a system of mode voltage vectors into a system of phase voltage vectors
φ	phase angle

Contents

1 Introduction 1
1.1 Standards and Regulatory Documents 2
1.2 Structure of DPLC Equipment 4
1.3 Component Parts of PLC Equipment 7
1.4 Programming, Control, and Access Functions 9
References 10

2 Modems of DPLC Equipment 11
2.1 DPLC Equipment as a Digital Transmission System: Shannon's Limit 11
2.2 Modulation and Noise Immunity Technologies 16
2.2.1 Modulation Technologies 17
2.2.2 Channel Encoding Methods 24
2.2.3 Methods for Compensating Distortions of the HF Path Frequency Response and Phase Frequency Response 30
2.3 DPLC Modems Bit Rate 33
2.4 Dynamic Bit Rate Adaptation 34
2.5 Time Delay in Transmitting Information 35
2.6 Peak Factor of Modulated Signal 36
2.7 Comparison of Single-Carrier and Multi-carrier Modulation Methods Applied to DPLC Systems 37
2.8 Wideband Modems Based on DMT Modulation 39
References 42

3 Multiplexers and Network Elements of DPLC Equipment 45
3.1 Multiplexers 45
3.1.1 Typical Schemes for Multiplexer Application in DPLC Equipment 45
3.1.2 Processing of Speech Signals. Low Bit Rate Speech Codecs 47
3.1.3 Prioritization of Information Signals 48
3.1.4 Transit of Information Signals 49

3.2 Network Elements 50
3.2.1 Structure of DPLC Equipment Network Element 51
3.2.2 Filtering and Prioritizing Network Traffic 53
3.2.3 Header Compression 54
3.2.4 Payload Compression of Data Packets 61
3.2.5 SNMP Monitoring of DPLC Equipment 62
References 63

4 Features of Frequency Characteristics of the HF Paths 65
4.1 Features of Attenuation Frequency Dependences of the HF Paths 67
4.1.1 Brief Information About Modal Parameters 68
4.1.2 Frequency Response of the Line Path Without Considering Multiple Reflected Waves 72
4.1.3 Effect of Reflected Waves 79
4.1.4 Effect of Weather Conditions 83
4.1.5 Effect of Switching of Power Lines 92
4.1.6 Frequency Response of the HF Path with HF Bridge 94
4.1.7 Features of Frequency Response of the HF Path Along Double-Circuit Overhead Lines 98
4.1.8 Frequency Response of the HF Path Along Cable and Cable-Overhead Power Lines 111
4.2 Return Loss and Input Impedance of the HF Paths 120
4.3 Characteristics of Disturbances in the HF Paths 128
4.3.1 Corona Noise 129
4.3.2 Transient Disturbances 141
4.3.3 Narrowband Interferences 141
References 142

5 DPLC Channel Design Issues 145
5.1 Main Tasks of the DPLC Channel Designing 145
5.2 Types and Characteristics of DPLC Information Traffic 146
5.3 Information Interaction in Electric Grid 148
5.4 Types of DPLC Channels 154
5.4.1 DPLC Channel with Time Division Multiplexing of Information Signals 155
5.4.2 DPLC Channel with Time Division Multiplexing of Information Signals and Packet Traffic Transmission 158
5.4.3 DPLC Channel with Packet Traffic Transmission 160
5.5 DPLC Channels with Transit Stations 165
5.5.1 The Simplest Model of DPLC Channel with Transit Stations 165
5.5.2 Limiting Transit Sections Based on the Delay and Speech Transmission Quality Criteria: Characteristics of Speech Codecs 166

5.6 Providing Reliability of DPLC Channels when Selecting Operating Frequencies 171
5.6.1 Calculation of the PLC Channel Overlapped Attenuation 172
5.6.2 Required Attenuation Margin 174
5.6.3 Application of Accurate Calculation Methods for HF Path Parameters 177
5.6.4 Providing EMC of APLC and DPLC Channels Operating in One Electric Grid 179
5.7 Design of DPLC Channels with Bit Rate Adaptation 181
5.7.1 Purposes of Use of DPLC Channels with Bit Rate Adaptation 181
5.7.2 Calculation of the PLC Channel Maximum Allowable Frequency 183
5.7.3 Long-Term Monitoring of DPLC Channels 198
References 199

Annex No. 1. General Description of WinTrakt and WinNoise HF Path Calculating Programs 201

Index 215

Chapter 1
Introduction

Theoretical basis of the signal propagation theory used in PLC communication was developed by J. R. Carson (1926, 1927). Significant contribution in theory of PLC technologies including methods for calculating and designing HF paths was made by J. Fallout, G. E. Adams, F.C. Krings, L.M. Wedephol, and C.H. Gary. Independently of them Russian scientists and engineers G.V. Mikutskiy, J.L. Bikhovskiy, M. V. Kostenko, L.S. Perelman, K.I. Kafeiva, V.S. Skitaltsev, and Y.P. Shkarin have made a lot of researches and developed the scientific basis for calculating HF path parameters, PLC channels design, and frequency planning. Most of the basic researches in PLC communications were conducted in the 1950s–1970s of the last century. Study of application of new PLC systems and design of digital networks using PLC equipment is described in publications of this book authors. Contemporary authors include M. Zajc, N. Suljanovic, A. Mujcic, R. Piggy, R. Raheli, E. Fortunatto, and A. Raviola.

Power line carrier communication channels are widely used for transmitting speech, data, teleprotection signals (relay protection (RP), and emergency automation (EA) commands). PLC channels are characterized by high mechanical strength which is determined by strength of transmission line used as the transmission medium, low cost due to the lack of costs for special linear structures, and effectiveness due to low operation costs. Let us recall that PLC frequency range in most countries of the world is limited to 40...500 kHz band; in Russia and the CIS countries, the frequency range used for PLC is within 16...1000 kHz

This book discusses time-division multiplexing (TDM) PLC communication systems used to build digital PLC (DPLC) communication channels intended for speech and data transmission. Various digital modulation methods are used in DPLC equipment: quadrature amplitude modulation (QAM), orthogonal frequency division multiplexing (OFDM), and discrete multi-tone modulation (DMT).

In modern electric power industry, DPLC equipment is used to build communication channels with information bit rate from several to hundreds of kilobits per second. In the last 20 years, developments of digital PLC system technology

A. G. Merkulov et al., *High Voltage Digital Power Line Carrier Channels*,
https://doi.org/10.1007/978-3-030-58365-1_1

periodically resort to the use of DMT modulation and DPLC channels with a nominal transmission/receiving frequency band hundreds of kilohertz. There is only small experience of practical application of such channels and, accordingly, no rules of their design (e.g., selection of frequencies). However, Chap. 2 will discuss features of DPLC systems with DMT modulation.

Today, according to rough estimates, about 250 thousand PLC channels are operated in the world. According to the Federal Grid Company of the Unified Energy System "FGC UES," number of PLC channels operated in Russia exceeds sixty thousand.

As for capital costs, low density of technological channels, and low noise, the use of digital PLC channels may be the most convenient way for creating communication between 35 and 110 kV substations. In the segment of 220 kV transmission lines equipped with optical groundwire and fiber-optic transmission systems, DPLC channels can be used as backup for voice and telecontrol. Use of DPLC systems for 330 and 500 kV lines is problematic. As a rule, these lines have a large attenuation and noise level. Nevertheless, there are examples of reliable DPLC channels for 500 kV lines with a length of 300 km...400 km operated in the lower PLCs frequency range (less than 100...150 kHz).

This publication was prepared with the information support of the Russian National Committee of CIGRE (RNC CIGRE). The authors thank the International Electrotechnical Commission (IEC) for permission to reproduce information from its International Standards.

1.1 Standards and Regulatory Documents

The first standard of the International Electrotechnical Commission, IEC 60663 [1], concerning usage of PLC equipment appeared in 1980. In 1993, IEC adopted standard No. 60495 "Single sideband power-line carrier terminals" [2]. The following standards were also issued for HF coupling equipment: IEC 60481 [3], IEC 60358-2 [4], and IEC 60353 [5].

CIGRE, International Council on Large Electric Systems, has issued two reports on the digital PLC channels:

- In 2000, report of working group 09 of 35 Study Committee: "SC35.9 CIGRE TB 164 Report on Digital Power Line Carrier" [6] defining basic principles of DPLC equipment designing.
- In 2006, report of working group 08 of D2 Study Committee (SC D2 is the new designation of RC 35): "D2.08 DPLC Present Use and Future Applications" [7] defining areas of application of DPLC equipment and covering general issues of design and commissioning of DPLC channels.

In 2010, IEC started updating standards 60663 and 60495 and preparing standards for application of PLC systems in medium-voltage (<35 kV) grids [8–10]. Information on new standards is given in Table 1.1. General information about design and operation of high voltage PLC channels is given in [11–13].

Table 1.1 IEC Standards defining requirements for PLC systems and PLC channel parameters

New standard	Old standard	Purpose and year of production
IEC 62488-1	IEC 60663	Planning of analogue and digital power line carrier systems operating in EHV/HV/MV electricity grids (2011)
IEC 62488-2 IEC 62488-3 IEC 62488-4	IEC 60495	Analogue power line carrier terminals (APLC) (2013). Digital power line carrier terminals (DPLC) (2019). Broadband power line systems (BPLC) for low and medium voltage (under development) IEC 62488-1 = Analogue....(APLC) IEC 62488-2 = Digital....(DPLC) IEC 62488-3 = Broadband....(BPLC)

As an example of regulation of PLC channels design and operation in an electric grid company, we give a brief overview of standards for organizations being a part of the Federal Grid company of Russia related to PLCs. These standards cover all main aspects of designing HF paths; they are used for selecting the operating frequencies of PLC channels, selecting couplings equipment; and they also provide typical solutions for creation PLC channels. Let us recall that there are more than 60 thousand PLC channels for various purposes in Russia, and, to avoid mutual interference with existing systems in a crowded frequency spectrum, it is very important to ensure high-quality design for new PLC channels.

1. STO 56947007-33.060.40.045-2010 – "Guidelines for selection of frequencies of high-frequency channels along 35, 110, 220, 330, 500 and 750 kV electric power transmission lines" [14]. This document includes information on how to select PLC channel operating frequencies, calculate attenuation of HF paths, and calculate mutual interference of HF systems. At that, such parameters of power transmission lines as voltage class, type of phase conductor, arrangement of conductors, weather conditions, etc. should be considered. Appendices to this document contain information about parameters of PLC equipment produced by various world manufacturers. In Chaps. 4 and 5 of this book, we will often refer to information provided in the guidelines.
2. STO 56947007-33.060.40. 052-2010 – "Guidelines for calculating parameters and selecting schemes of high-frequency paths on 35...750 kV AC power transmission lines. This document includes information about methods for calculation of HF path frequency response, for choosing optimal HF coupling schemes which allow ensuring the best performance of the PLC channel".
3. STO 56947007-33.060.40.108-2011 – "Norms for designing PLC systems." The standard contains information on scope of documents, diagrams, and drawings which should be included in the PLC channel project.
4. STO 56947007-33.060.40.125-2012 – "General technical requirements to line traps, coupling capacitors and coupling devices for PLC channels along 35... 750 kV electric power transmission lines." This standard contains typical requirements to various technical parameters of line traps (LT), coupling capacitors (CC) and coupling devices (CD), for example, line trap impedance, capacitance of the coupling capacitor, input impedance of coupling device, etc. Coupling devices also called as line matching units (LMU).

5. STO 56947007-33. 060. 40. 178-2014 – "Technological communication. Operation manual on PLC communication channels along 35... 750 kV electric power transmission lines." This document is an instruction on how to operate PLC channels: to measure frequency response of HF paths, line traps, and coupling devices, to perform technological maintenance of PLC equipment.

1.2 Structure of DPLC Equipment

Block diagram of DPLC equipment, excluding teleprotection described in IEC62488-3 standard and CIGRE reports, is shown in Fig. 1.1 (without reference to a specific type of system).

Voice frequency (VF) modem interfaces (A), analog voice interfaces (B), and voice signaling interfaces (C) are in the voice-frequency interface unit. Signals from the latter are sent to digitization and compression unit. Digital voice interfaces (V) are also connected to this unit. RS-232/485, X. 21/V. 11 serial data interfaces (S) are in the multiplexer with time division of signals, and Ethernet interfaces (T) are in the network element, for example, Ethernet switch. The digitization and compression unit is also connected to the multiplexer via an internal bus. The multiplexer unit is connected to the digital modem unit (Y). For transmission, the modem generates a high-frequency signal which enters line HF signal processing unit via the modem line interface (G1). Here, HF signal is amplified, filtered, and matched for transmission to communication line via line HF interface (G2). Input/received HF signal enters the DPLC system via the same line interface (G2), which is filtered and transmitted to the modem. The diagram also shows connection points for parameterization and maintenance (M) interface and power supply ports (P). When operating in mixed analog–digital mode, ADPLC equipment block diagram becomes more complex and has the form shown in Fig. 1.2.

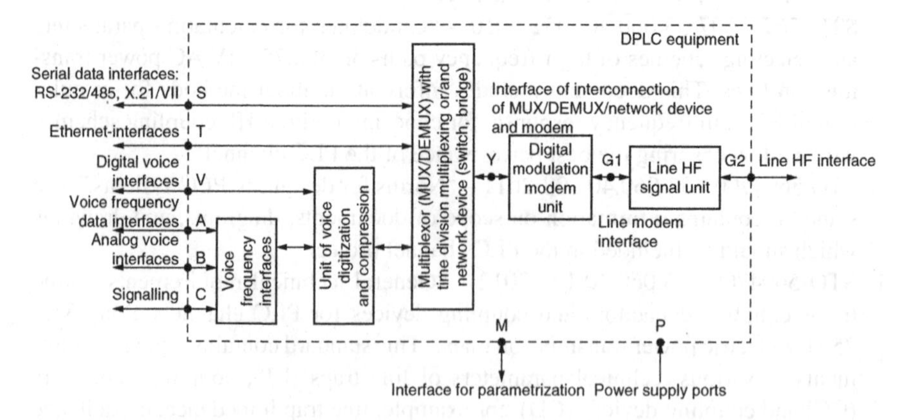

Fig. 1.1 DPLC equipment block diagram

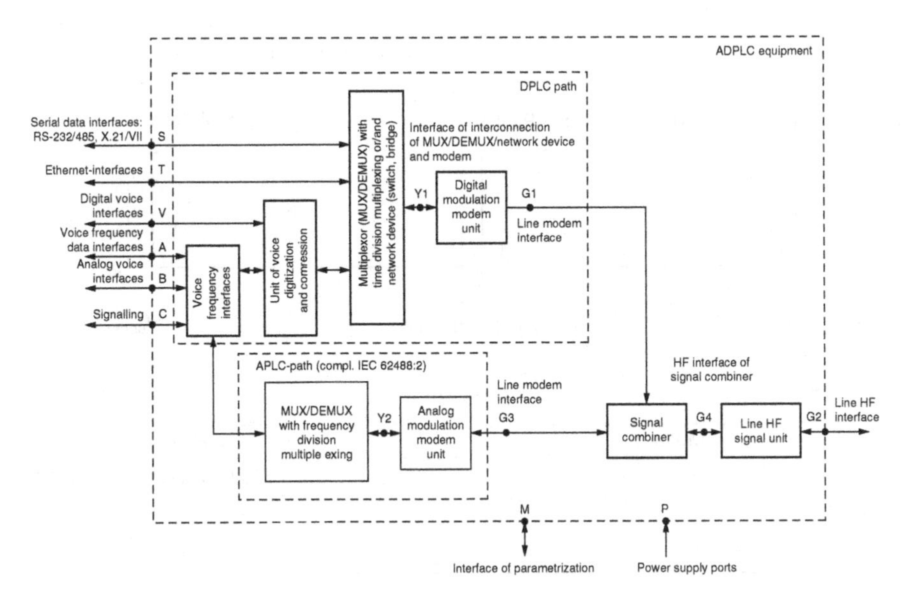

Fig. 1.2 ADPLC equipment block diagram

APLC unit is added to the scheme; it is an analog part which complies with requirements of IEC62488-2 standard and includes a multiplexer with frequency division of signals. Analog voice and data transmission signals from VF interface unit can be distributed either to the digitizing and compression unit of DPLC subsystem or to the multiplexer of APLC unit which sends signal to amplitude modulator with a single side band (SSB) modulation. Signals from modem units (G1) and (G3) are summed in HF signal combiner unit (G4) and then are sent to line HF signal processing unit (G2).

According to the method of forming a group HF signal, the equipment can be single channel or multichannel.

In the first case, nominal transmission bandwidth of one channel – $n \times 4$ kHz – can be flexibly distributed among all information services. The system operation is controlled by a single system pilot signal.

In the second case, nominal transmission band is divided into $n = 4$ kHz subchannels, and each subchannel has its own system pilot signal.

PLC equipment pilot signals are used for automatic gain control (AGC) and receiver clock synchronization, channel parameters measurement, and they can be also used for transmitting voice signaling of voice channel or transmitting internal service information.

DPLC systems can also be divided into two types:

- Entire frequency band of $n \times 4$ kHz channel is used for operation of a single modem. In such systems, one common digital modulated signal is generated.
- In separate frequency bands, such as 4, 8, or 12 kHz, own modem is used. In this case, total system bit rate is equal to the sum of bit rates of all modems in all frequency bands.

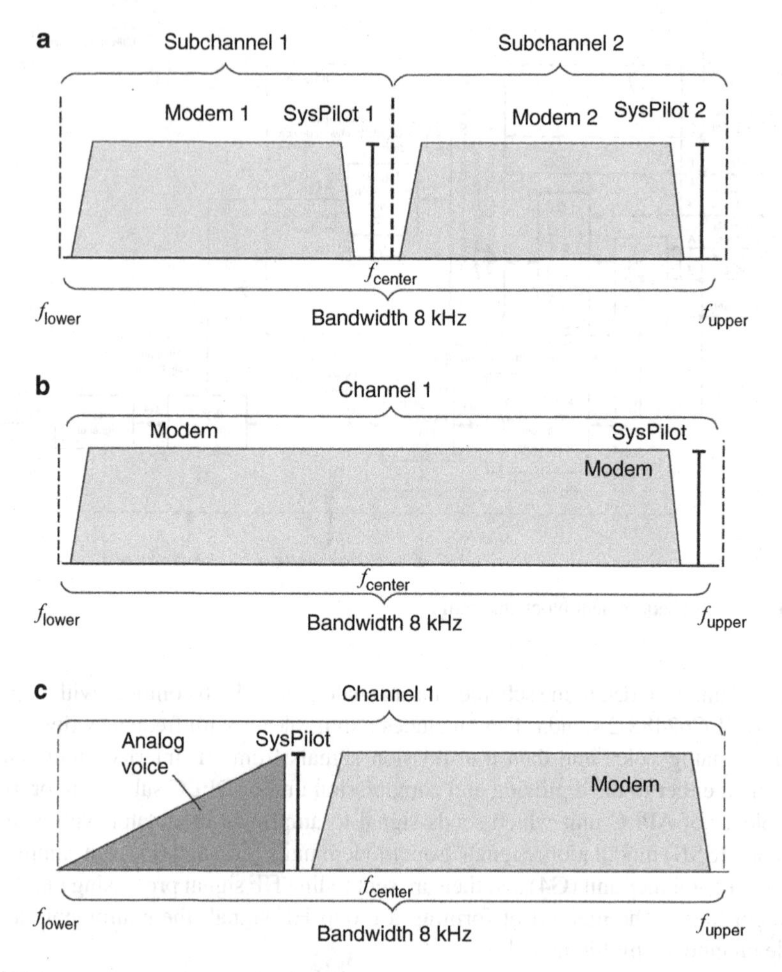

Fig. 1.3 Examples of using nominal frequency band in DPLC and ADPLC systems

Figure 1.3a, b shows examples of using nominal bandwidth of DPLC equipment for two operation options: the one with multiple modems and the one with a single modem.

Example of bandwidth utilization for ADPLC system is shown in Fig. 1.3c. In this case, a part of the frequency band is intended for APLC services, and the remaining part is used for DPLC. There can be one pilot signal, as shown in the figure, or there can be several pilot signals where each channel has its own signal. In practice, various schemes for dividing and forming a group HF signal are used.

1.3 Component Parts of PLC Equipment

In most cases, PLC equipment consists of two sections located on different chassis: line section of HF signal processing and section of central processor unit, which includes all units for digital processing of DPLC and APLC signals. There are systems where both sections are located on the same chassis. An example of PLC equipment component diagram is shown in Fig. 1.4. Section of central processor includes the following modules.

1. Central processor unit (CPU) module.

The CPU module performs all mathematical operations for digital signal processing (DSP): signal conversion, encoding, modulation, demodulation, and decoding, as well as working with HF signals.

To do this, the CPU module board has microcontrollers, processors, and FPGAs (field-programmable gate array) of multiplexers with time and frequency division of signals and modems with digital and analog modulation. Analog-to-digital and digital-to-analog converters (ADC and DAC), clock frequency oscillators, nonvolatile memory, AGC circuits, attenuators (ATT), and amplifiers are also included in the CPU module.

Digital data interfaces, network switches can also be located here; all these options are the design features of equipment from different manufacturers.

The CPU module contains interfaces for PLC equipment parameterization, measuring connectors, and light signaling elements.

In some PLC equipment samples, a licensed function electronic key which contains information about number and type of services and subscriber terminals available for operation and configuration is located on the CPU module board.

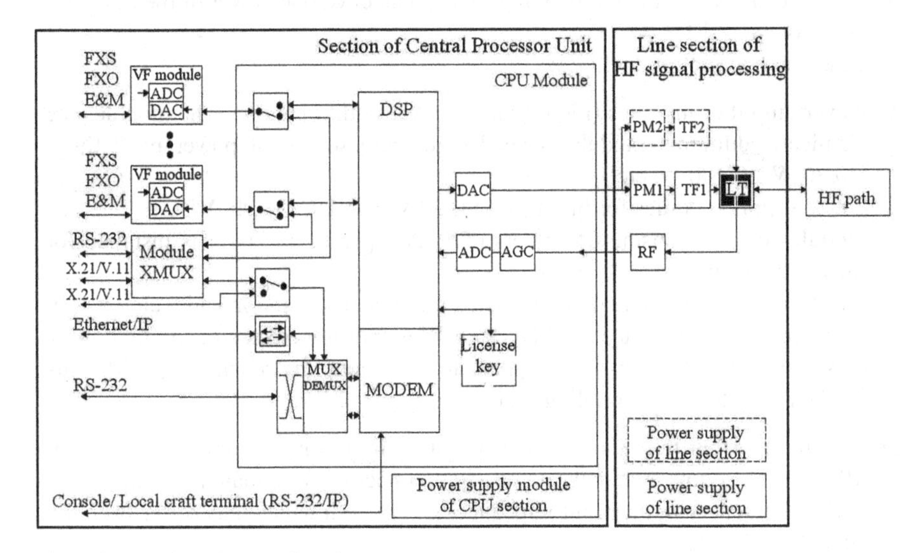

Fig. 1.4 Diagram of the PLC equipment component parts

2. Modules for voice frequency analog interfaces.

Such voice interfaces as FXS, FXO, E&M, and signaling and 4-wire voice-frequency interfaces for connecting analog modems are located here. All modules include ADC and DAC, so data exchange with the central processor via internal bus is carried out using digital signals.

3. Multiplexer module (XMUX, optional).

When a multiplexer is implemented as a separate module, its capabilities can be significantly expanded compared to a multiplexer embedded in a central processing unit (MUX). This is mainly applied to speech signal compression functions when using various low bit rate speech codecs, priority management of information signals, and their routing.

4. Power supply module for CPU section.

The module is used for generating supply and control voltage for all modules of the CPU section.

Connections of different modules inside the CPU section are performed via different data buses and HF interfaces.

Line part of the equipment is a set of modules designed for transmitting HF signal to the communication line and, accordingly, receiving HF signal from the communication line.

Transmitted HF signal passes from modem output through power amplifiers, line filters of the transmitter, and is transmitted to the line through a line matching transformer. From the amplifier, the signal is transmitted to the line filter through internal matching transformer. The amplifier operation can be controlled by the CPU module via internal interface circuits. In most up-to-date DPLC systems, nominal output power of the transmitter corresponding to the peak envelope power of the transmitter does not exceed 100 W.

Thus, the line section includes:

1. Power amplifier modules (PM). Main one and additional one. The modules are identical; additional module is needed to increase output power by 2 times (40/80 W, 50/100 W, etc.).
2. Power supply modules for line section (220 VAC/VDC or −48 VDC).
3. Modules of line transmission filters (TF). A separate line filter is installed for each power amplifier.
4. Line transformer (LT) module. It is used for galvanic isolation and matching HF interface of the equipment and HF path. Nominal impedance of the line transformer output is 75 ohms for the phase-to-earth coupling scheme and 150 ohms for the phase-to-phase coupling scheme.

Manufacturers often introduce additional output impedance nominal, such as 25, 50, 100, and 125 ohms, that makes it possible, if necessary, to match the line output of PLC equipment with the input impedance of HF path which may differ from the nominal value 75 (150) ohms.

The line transformer is also needed to protect the amplifier against external interfering signals and pulse interference coming to the equipment from the coupling device (line matching unit).

5. Receiving HF signal, from the matching line transformer, enters the receiver filter and then enters the modem receiver input.

Receiver filter module (RF). It includes attenuators and amplifiers in addition to filter elements.

There are matched receivers with a nominal impedance of 75 ohms for phase-to-earth coupling and 150 ohms for phase-to-phase coupling. Some PLC systems use unmatched receivers with an input impedance of more than 1 kiloohm. A high-impedance unmatched receiver is easier to implement in practice, but it adds additional non-uniformity to HF path attenuation characteristic and, in some cases, worsens return loss on the transmitter side.

The receiver filter provides preliminary filtering; highly selective filtering in the CPU module is performed using digital signal processing. In addition, in the line part of the equipment, when operating in the adjacent frequency mode, special differential transformers designed to suppress signal of own transmitter in the direction of its own receiver can be used.

1.4 Programming, Control, and Access Functions

As a rule, from the point of view of software implementation, modern PLC systems have several levels.

1. Primary firmware is installed by manufacturer during the hardware production: it is like the BIOS for a computer.
2. Main firmware is a program which controls all hardware units. It can be changed when manufacturer adjusts the program code to fix errors or add new algorithms and functions.
3. Hardware configuration software. Software which includes operator interface and is used to configure hardware.
4. Firmware for network elements. In some systems, DPLC equipment network element is programmed separately via a web browser, but not via configuration software. To change firmware of network elements, command line interface (CLI) and FTP server are used.

During operation, central processor of DPLC equipment constantly tests all the system units for software and hardware errors. When problems are detected, the equipment generates warning and alarm signals to LEDs and alarm relays.

Access to hardware is provided via local or remote connection. Local connection to the equipment is made either via the console port and USB, RS-232 interface, or a TCP/IP connection and Ethernet network. Access to a PLC device is protected by password and user account, such as user/specialist/administrator. Remote access to

PLC terminals can be implemented via in-band monitoring channel. Thus, being on the one side of the channel, operator can access the equipment located on the opposite side by means of service signals.

As mentioned above, there are examples of PLC systems in which hardware functionality available for use is defined in the electronic license key. For example, multiplexer can process ten voice signals, but only three licenses are open for use in an electronic key. The same principle is used to activate data transfer interfaces, remote monitoring channels, and so on. Here a flash memory with a unique serial number serves as a key [15]. The key memory contains information about number and type of available functions and interfaces. The key is located on the board of CPU module. When ordering hardware function extensions, manufacturer either generates a special code which is loaded into the key memory via PLC device system maintenance program or provides a new electronic key which is to be installed in the hardware instead of the old one.

References

1. IEC 60663 Planning on (single-sideband) power line carrier systems. Technical report, IEC (1980)
2. IEC 60495 Single sideband power-line carrier terminals. International Standard, IEC (1993)
3. IEC 60481 Coupling devices for power line carrier systems. International Standard, IEC (1974)
4. IEC 60358-2 Coupling capacitors and capacitor dividers - Part 2: AC or DC single-phase coupling capacitor connected between line and ground for power line carrier-frequency (PLC) application. International standard, IEC (2013)
5. IEC 60353 Line traps for a.c. power systems. International standard, IEC (1989)
6. SC35.9 CIGRE TB 164 Report on digital power line carrier. Technical brochure. CIGRE WG 35.09 (2000)
7. D2.08 DPLC Present use and future applications. Technical brochure, CIGRE TF D2.08 (2006)
8. IEC 62488-1 Power line communication systems for power utility applications – Part I. Planning of analogue and digital power line carrier systems operating over EHV/HV/MV electricity grids. International Standard, IEC (2011)
9. IEC 62488-2 Power line communication systems for power utility applications – Part II. Analogue power line carrier terminals or APLC. International Standard, IEC (2018)
10. IEC 62488-3 Power line communication systems for power utility applications – Part III. Digital power line carrier terminals or DPLC and hybrid ADPLC terminals. International Standard, IEC (2019)
11. Bulletin PLC 79-1 The PLC Handbook, Dowty RFL Industries Inc., US (1979)
12. REA Bulletin 66-5: Power System Communications: Power Line Carrier and Insulated Static Wire Systems, pp. 78–22742. Energy Management and Utilization Division Rural Electrification Administration, U.S., Government Publications (1978)
13. Sanders, M.P.: Power line carrier channel & application considerations for transmission line relaying. In: Sanders, M.P. Pulsar Technologies (1997)
14. Standard of Organization: Guidelines for Selection of Frequencies of High-Frequency Channels Along 35, 110, 220, 330, 500 and 750 kV Electric Power Transmission Lines. Federal Grid Company of the Unified Energy System of Russia, Moscow (2010)
15. Dallas Semiconductor Book of DS19xx iButton Standards, US https://pdfserv.maximintegrated.com/en/an/AN937.pdf (1995). Accessed 01 June 2020

Chapter 2
Modems of DPLC Equipment

2.1 DPLC Equipment as a Digital Transmission System: Shannon's Limit

Theoretical foundations of digital communication were laid in the 20–30s of the twentieth century by R. Hartley [10], G. Nyquist [22], K. Shannon [27], and V. A. Kotel'nikov [15] who proved the basic theorems of information theory. Modern digital transmission systems are well-described in the books by B. Sklyar, D. Prokis, and L. Hanzo [9, 26, 28].

Figure 2.1 shows structure of DPLC equipment as a classic digital transmission system. Detailed information about the principles of DTS design and technologies is presented in [9].

Such digital transmission system includes information sources, source encoders, multiplexer, channel encoder, modulator, communication line, channel decoder, demultiplexer, source decoders, and information receivers.

Various devices for collecting and transmitting information which output and input signal has a digital form (Fig. 2.2) and use two-level or three-level encoding at the physical level can also be sources and receivers of the signal. Examples are RS-232, E1, and Ethernet digital interfaces. The following codes are applied: NRZ (No Return to Zero), RZ (Return to Zero), Manchester encoding, HDB3 (high-density bipolar code), etc.

Multiplexer combines several incoming information signals into one aggregated signal intended for channel encoding in the encoder. Demultiplexer performs the reverse conversion, accordingly.

Channel encoder and decoder are used to increase noise immunity of the DPLC. When decoding information, error correcting encoding allows detecting and correction of a certain number of errors by adding redundancy to the source code combination.

A. G. Merkulov et al., *High Voltage Digital Power Line Carrier Channels*,
https://doi.org/10.1007/978-3-030-58365-1_2

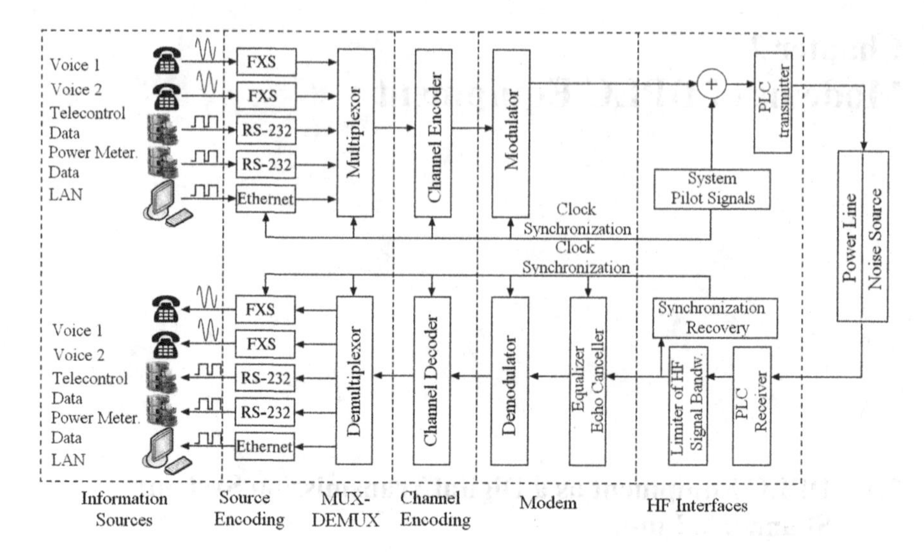

Fig. 2.1 DPLC equipment as a digital transmission system

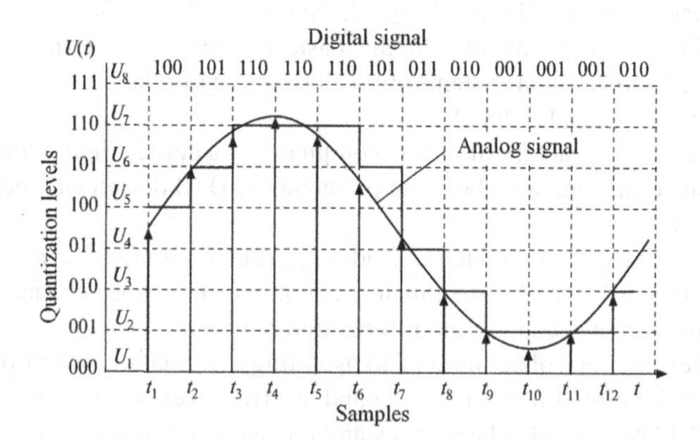

Fig. 2.2 Sampling and quantization of analog signal

Adding redundancy to the original message inevitably increases requirements to DPLC bit rate. For example, when using code with an encoding rate of 1/2, each bit of the original data code sequence is encoded with two bits, and, accordingly, amount of data transmitted to the line is doubled, and bit rate required is doubled.

Error correcting codes are divided into two main classes: continuous and block codes. Continuous convolutional codes perform sequential processing of small fragments of input information. Block cyclic codes do not operate with individual bits, but with blocks, i.e., groups of bits, some codes operate with bytes (octets) of information. Convolutional codes are good at correcting single errors caused by noise, but they are not good at dealing with error packets which occur, for example, due to pulse interference. Cyclic codes, on the other hand, can correct packets of errors with a size which depends on the size of encoded data block. Turbo codes are the most advanced in terms of providing noise immunity. Nevertheless, they are very complex; they cause additional delay of transmission, work well with a high probability of error, but are less effective if the probability of error is low.

Modulator converts digital code sequence into analog high-frequency signal to transmit the signal through communication line. Demodulator performs reverse conversion, accordingly.

Communication line is a medium for transmitting information signal. Communication line contains noise and non-homogeneities of various natures that worsen the conditions for information receiving. Features of electric power transmission lines as a medium for transmitting HF signal will be discussed in Chap. 4.

For DTS to work with specified bit rate and probability of erroneous reception in communication lines which have small non-uniformity of frequency-phase characteristics, two conditions should be met:

- Sufficient signal-to-noise ratio (SNR) in the receiver
- Absence of inter-symbol interference due to the influence of individual information elements (symbols) on each other

SNR (signal-to-noise ratio), in logarithmic representation, shows difference between received signal level p_{RX}, dBm, and noise level p_{noise}, dBm, in the modem operation frequency band (Fig. 2.3). Level of received signal, in turn, is equal to the difference of level of transmitted signal p_{TX}, dBm, and its attenuation after passing through the communication path a_{path}, dB.

Dynamic changes in HF path parameters lead to changes in SNR at the receiver input – one of the main parameters which affect modem operation in any digital transmission systems. This is discussed in Chap. 5.

The more is SNR, the better conditions for receipt of information are. When SNR decreases, probability of receiving incorrect data which is defined by bit error rate (BER) and block error rate (BLER) increases.

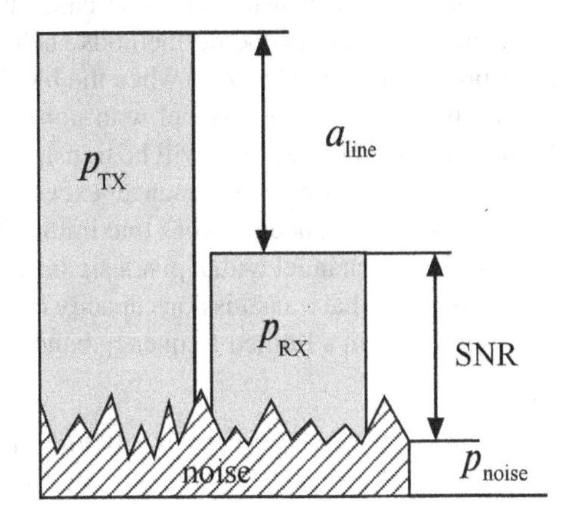

Fig. 2.3 Signal levels, SNR

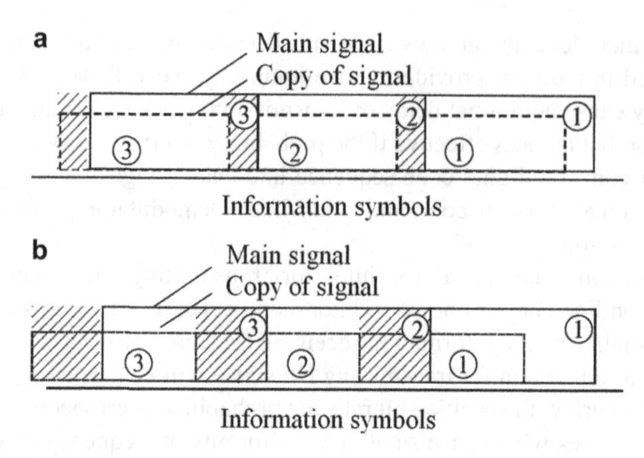

Fig. 2.4 Inter-symbol interference: (**a**) weak inter-symbol interference; (**b**) strong inter-symbol interference

Inter-symbol interference (ISI) is a superimposition of consecutive information symbols on each other. Main reason for ISI occurrence is a distortion of the HF path frequency response characteristic and group time delay in the channel frequency band. It can be caused by multipath propagation of the signal, in which a copy of information signal arrives at receiver's input with some delay due to, for example, reflections from non-homogeneities in the communication line. In this case, some information symbols become obstacles to receipt of other symbols. Figure 2.4 gives an example of inter-symbol interference with varying degrees of distortion.

There are more causes for ISI occurrence in DPLC channels. They can be, for example, power line transpositions, tap lines, power lines running parallel, etc. that form non-homogeneities and cause multiple reflections and distortion of HF path transmission characteristics and, consequently, distortion of phase and amplitude of received HF signal. In any case, receiving and processing input signal in DPLC is much more complex than in other types of communication.

According to the Shannon–Hartley theorem [27], if performance of information source $R_{\inf}$ is less than transmission capacity of communication channel B, then there are codes and decoding methods such that average and maximum decoding error probabilities tend to zero when the block length tends to infinity.

In other words, for a channel with noise, such encoding system can be always found using which messages will be transmitted with arbitrarily high degree of reliability, unless source performance exceeds the channel transmission capacity. Shannon's limit is understood as maximum data bit rate for which error correction is possible in a channel with a given signal-to-noise ratio.

Let us recall that transmission capacity of a channel with additive white Gaussian noise (AWGN) in a limited frequency band is defined as [27]:

$$B = W \log_2 \left(1 + \frac{P_S}{P_N} \right)$$

where B channel transmission capacity, bit/s; W channel bandwidth, Hz; P_S signal power, W; and P_N noise power, W.

Spectral efficiency η which shows how many bits of information falls in 1 Hz (bit/s/Hz) of the channel bandwidth, per 1 second, is defined as:

$$\eta = \frac{B}{W} = \log_2\left(1 + \frac{P_S}{P_N}\right)$$

For DPLC channels, due to frequency dependence of signal and noise levels, the following formula should be used to define maximum transmission capacity:

$$B = \int_0^W \log_2\left(1 + \frac{P_S(f)}{P_N(f)}\right) df$$

When transmitting information, some energy of the transmitter E_b, W·s/bit, falls at one transmitted bit. At the same time, the communication channel contains noise with a certain spectral density P_{N0}, W/Hz. K. Shannon showed that there is a lower limit value of the ratio E_b/P_{N0} at which, at any information bit rate, it is impossible to carry out error-free transmission if ratio of energy E_b used for transmission of one bit of information to spectral noise density $N0$ is equal to −1.59 dB:

$$\frac{E_b}{P_{N0}} \geq \lim_{\eta \to 0} \frac{2^\eta - 1}{\eta} = -1.59[dB],$$

Shannon's limit can be illustrated by the graph shown in Fig. 2.5 [27]. Existing communication systems work below the curve $\eta(E_b/P_{N0})$; for them $R_{inf} < B$. Curve $\eta(E_b/P_{N0})$ defines Shannon's limit for a given value of E_b/P_{N0}. The area above the curve E_b/P_{N0} is inaccessible.

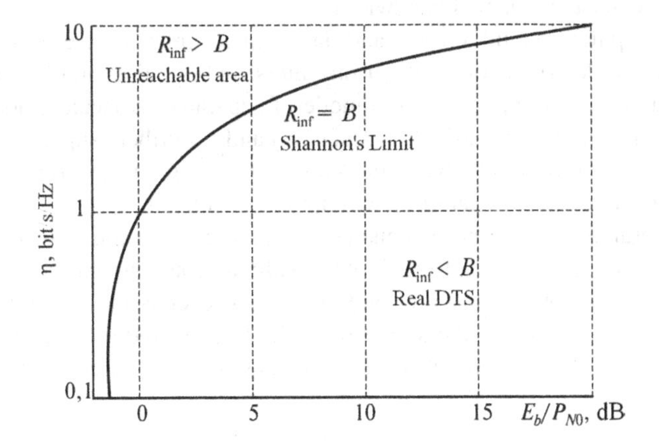

Fig. 2.5 Shannon's limit

As will be shown in Chap. 4, noise in high voltage electric power transmission lines is represented by several sources of different nature. For analog HF systems, on the not damaged transmission lines with voltage less than 110 kV, it may be represented as AWGN. For digital PLC systems, correction factor depending on voltage class of the overhead line should be considered. There is a corona noise of phase conductors of power lines with a voltage of 110 kV or more. In comparison with channels affected by AWGN, when exposed to corona noise, 2... 9 dB higher SNR will be required for the same spectral efficiency of the modem. This issue will be discussed in Chap. 5.

2.2 Modulation and Noise Immunity Technologies

Modems are the most complex elements of DPLC equipment. Bit rate, delay, and noise immunity of information transmitted through the power line depends on error correcting codes and modulation methods used in the modem. Modems are characterized by the following parameters:

- Modulation technologies
- Channel encoding methods
- Data bit rate
- Data bit rate adaptation
- Information transmitting time delay (latency)
- Peak factor depending on modulation type

Building DPLC system modem is a solution of optimization problem to ensure required data bit rate for a given variation range of SNR, attenuation, distortion, and modes of transmission and required values of error rate and delay of information processing. As a rule, we can provide either high data bit rate with low delay and some acceptable SNR reduction or low rate with significant SNR reduction and large delay at constant channel bandwidth.

DPLC equipment includes voice and data transfer interfaces, encoder/decoder pairs, modulator/demodulator pairs, various filters, and equalizer and echo canceller.

PLC equipment can operate in the mode of separated, adjacent, and superimposed receiving and transmission frequency bands. Further, superimposed frequency band operation mode, when the same nominal frequency band is used for transmitting and receiving operation, will not be considered.

When operating at adjacent frequencies, nominal transmission and receiving frequency bands are located next to each other without protecting gap between them. At that, signal from own transmitter partially enters receiver of the equipment that makes it difficult to receive the signal from the line. To reduce impact of signal transmission on the receiver, special hybrid transformers and echo canceller are used in modems.

2.2.1 Modulation Technologies

Various types of modulation are used in digital transmission systems. The most common are single-carrier schemes: phase-shift keying (PSK), quadrature amplitude modulation (QAM), and multi-carrier schemes – orthogonal frequency division multiplexing (OFDM), and discrete multi-tone modulation (DMT). In contrast to QAM, when OFDM modulation is used, information is transmitted in parallel via a set of carrier frequencies. When DMT modulation is used, carrier frequencies are divided into groups and each group of carrier frequencies can use its own modulation scheme. One of advantages of multi-carrier modulation schemes is its ability to disable part of the carrier transmission/reception frequencies. This approach allows notching certain frequency ranges of DPLC channel operating frequency band occupied by other PLC channels or radio stations.

Here are some historical facts. Principles of QAM modulation were first described by C. Cahn in 1959 [3]. OFDM-modulation scheme was proposed by R.W. Chang in 1960 [5]. Digital modulation was rapidly developed thanks to S. Weinstein, P. Ebert, B. Hirosaki, I. Kalet, and others [12, 14, 31].

In the classical interpretation, modulation is the process of transferring a low-frequency signal f_1 to the high-frequency area by carrier frequency modulating f_2. In digital modulation methods, the carrier frequency is modulated by incoming code sequence. That is, n bits of information correspond to some analog signal with given amplitude A and phase φ which can be represented as follows:

$$S(t) = A(t)\sin\left(2\pi f_c t + \phi(t)\right)$$

The modulation is characterized by a modulation index M equal to 2^n and defining number of variations in the amplitude and phase of the modulated signal. For convenience, this set of variations is usually displayed on the complex plane as points of the signal constellation diagram.

Let us consider two the most common types of modulation widely used in DTS are phase-shift keying and quadrature amplitude modulation.

Phase-shift keying. The principle of phase-shift keying is to transmit symbols by affecting the phase of the modulated signal. In this case, the signal amplitude remains constant, only phase changes:

$$s_m(t) = A(t)\sin\left(2\pi f_c t + \phi_m(t)\right)$$

As examples, polar diagrams of signal constellations for two-position BPSK, four-position 4PSK, and eight-position 8PSK phase-shift keying are shown in the left column of Table 2.1.

Quadrature amplitude modulation. In this case, both phase and amplitude of the signal change, so that, in general, the channel symbol of M-level QAM modulation can be represented in a complex form as follows:

Table 2.1 Signal constellations for PSK and QAM modulations

Polar diagram of signal constellations for M-PSK modulation	Quadrature diagrams of signal constellations for M-QAM modulation
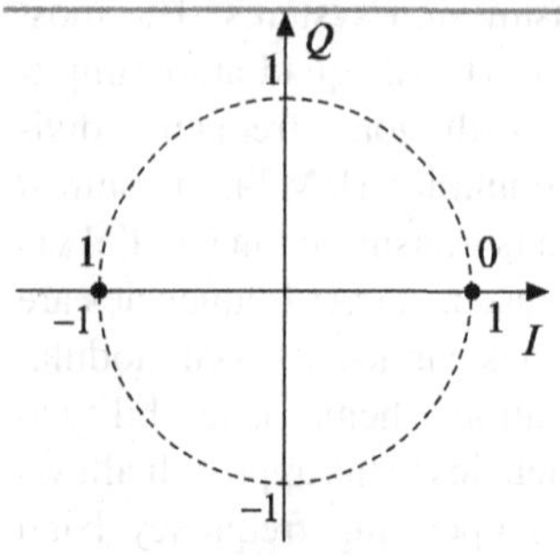 BPSK (binary phase-shift keying)	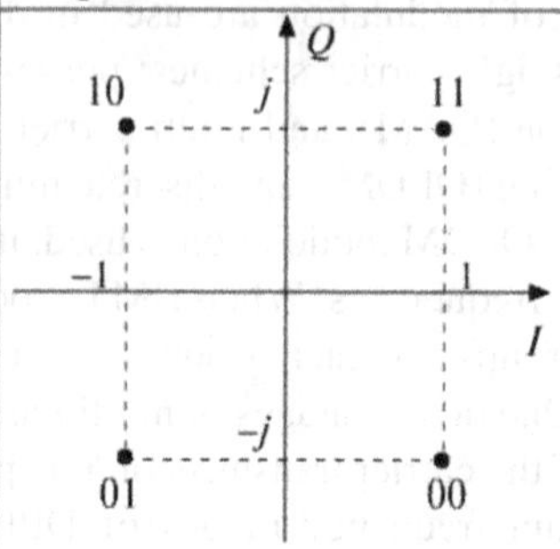 4QAM
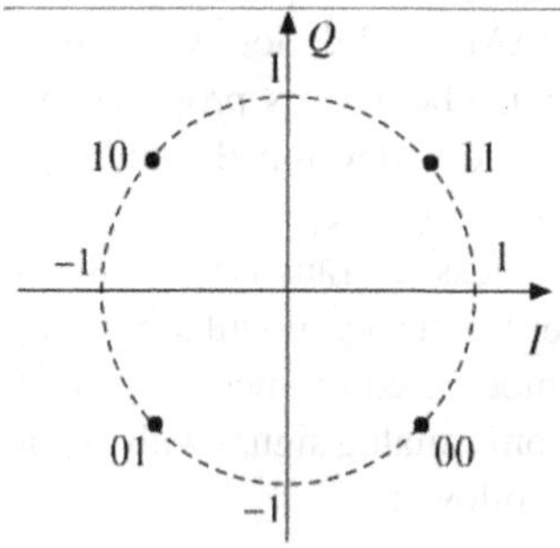 4PSK (QPSK)	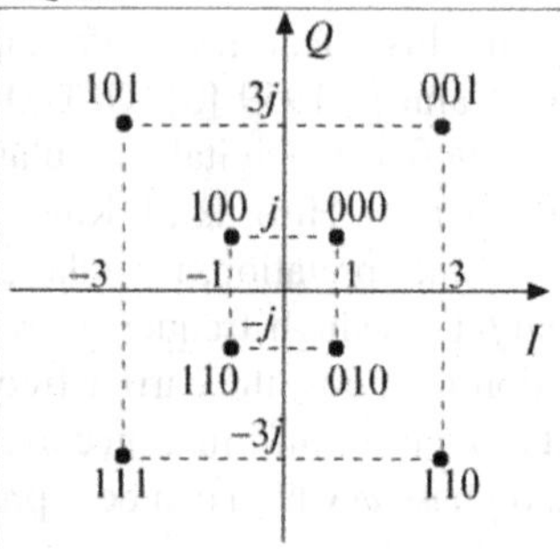 8QAM
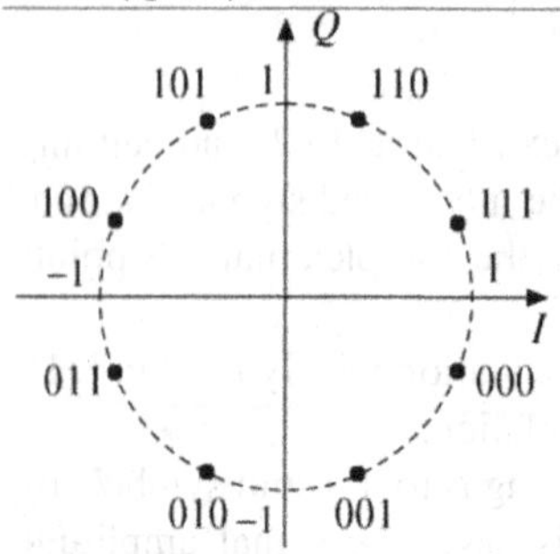 8PSK	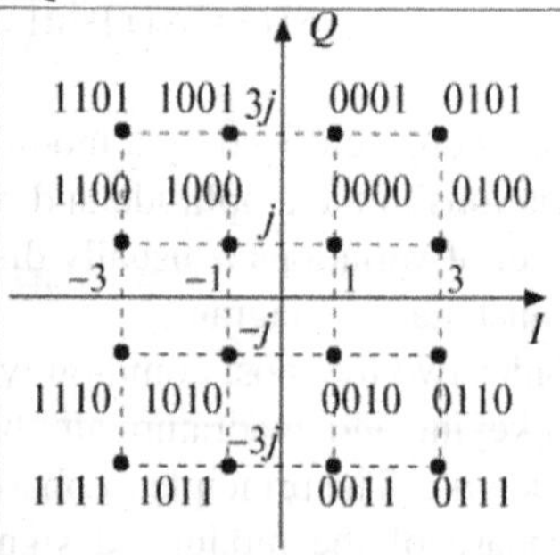 16QAM

$$s_m(t) = A_m(t) e^{j2\pi f_c t + \phi_m(t)}$$

The QAM modulation signal is the sum of two signals of the same frequency shifted relative to each other by 90° in-phase I and quadrature Q components, where a_m and b_m are coordinates of m-th point in the signal constellation:

$$s_m(t) = a_m \cos(2\pi f_c t) + j b_m \sin(2\pi f_c t)$$

Examples of signal constellations quadrature diagrams for 4QAM, 8QAM, and 16QAM are shown in the right column of Table 2.1.

Noise immunity of a modulation scheme depends on minimum distance between any two points of the signal constellation (Euclidean distance is the vector between two points of the signal constellation). The smaller the distance, the stricter the requirements to amplitude-frequency, phase-frequency, and noise characteristics of the communication channel and components of the receiving and transmitting equipment. At the same symbol rate (number of symbols transmitted per time unit), it is usually sought to apply modulation schemes with the largest Euclidean distance.

Figures 2.6 and 2.7 present graphs of typical spectral efficiency for PSK and QAM modulation with different M indices. In comparison with PSK with high modulation indices, $M = 16$ or more, QAM modulation gives a significant energy gain: lower SNR value is required to transmit information with a given bit error rate than for PSK. For example, in comparison with 16PSK, 16QAM gives gain in SNR of 4 dB, and for 64QAM in comparison with 64PSK, it is almost 10 dB.

Figure 2.8 shows block diagram of QAM modulator. This scheme can also be used to generate a PSK signal when transmitting information only using in-phase component.

DPLC systems do not require the use of intermediate frequency units and analog adders, which are necessary, for example, in radio communication systems operating in the range of hundreds and thousands of MHz. In PLC range, all conversions of low frequency signal are performed mathematically. Characteristics of modern DAC and ADC devices allow processing the signal in high-frequency range. An analog HF signal is generated using direct digital synthesis (DDS, direct digital synthesis) technology [21] which allows almost instantaneous changes in the signal phase and amplitude. Modulator circuit includes converter from serial to parallel signal, carrier frequency oscillator, and phase accumulator (accumulating adder used to generate phase code), look-up tables of the sin/cos carrier frequency, mixers, adder, DAC, and low-pass filter.

Initially, input serial bits sequence x is divided into two parallel sequences which modulate in-phase $Y_I = a_m \cos(2\pi f_c t)$ and quadrature $Y_Q = b_m \sin(2\pi f_c t)$ components.

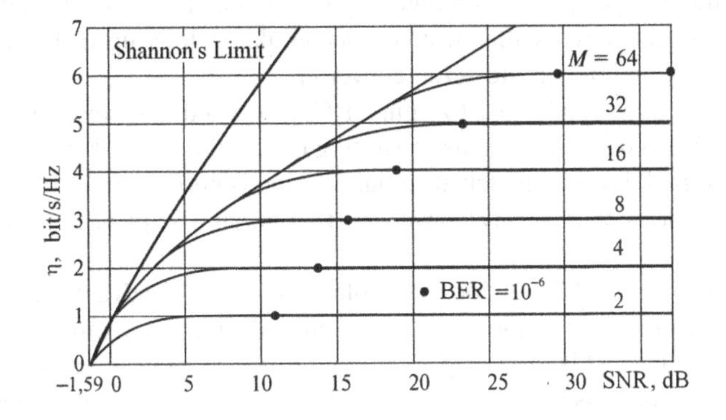

Fig. 2.6 Spectral efficiency of M-PSK modulation

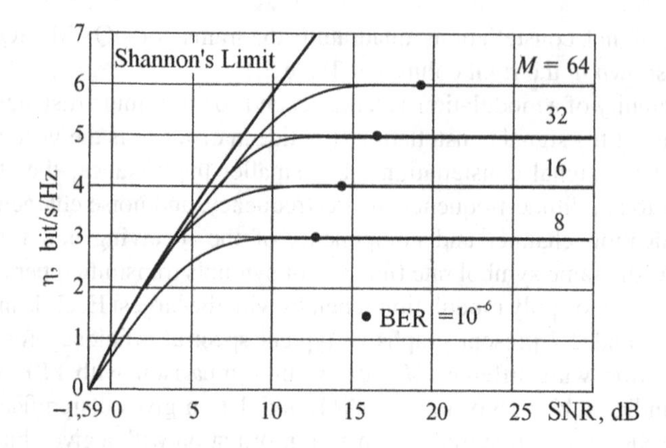

Fig. 2.7 Spectral efficiency of M-QAM modulation

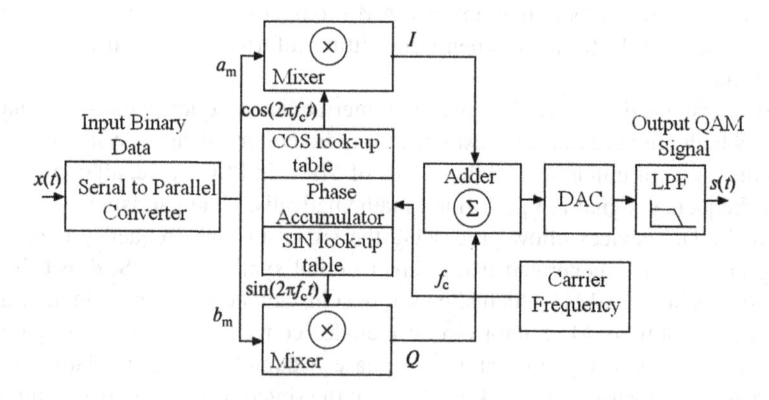

Fig. 2.8 Block diagram of QAM modulator

After summing these components, a signal is formed which is converted to analog signal, filtered, and transmitted as $s(t)$ to the communication line.

QAM demodulator diagram is shown in Fig. 2.9. It includes mixers and low-pass filters, carrier frequency oscillator, phase accumulator, look-up tables, analog-to-digital converter, and parallel-to-serial converter.

At reception, $s(t)$ signal is digitized in ADC, processed in mixers with multiplying by $\cos(2\pi f_c t)$ and $\sin(2\pi f_c t)$, thus separating the quadrature and in-phase components of the signal. Further, matched signal filtering is performed.

Original signal $x(t)$ is get by converting two parallel sequences to one binary sequence.

QAM modulation, while obviously simple, has several disadvantages when used in DPLC systems. For example, duration of one information symbol is defined as $t_{symbol} = 1/\Delta f$, where Δf is operating bandwidth of the modem. It can be seen from the formula that the pulse duration does not depend on modulation scheme, but the wider the frequency band, the smaller t_{symbol} value. And the less symbol duration, the

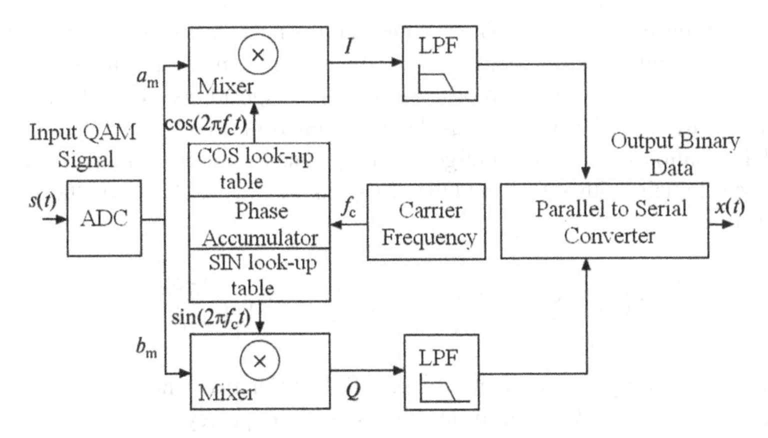

Fig. 2.9 Block diagram of QAM demodulator

worse conditions for its reception due to interference, abrupt changes in attenuation, and inter-symbol distortions in the presence of non-uniformity in the transmission characteristic of the communication line, etc.

OFDM modulation. Frequency multiplexing of carriers (FDM, frequency division multiplexing) is used in nonsensitive to these phenomena modulation schemes with multiple carrier frequencies. In this case, N separate frequency subchannels are formed on N carrier frequencies in the modem's frequency band. In DPLC technologies, the most common is multiplexing with orthogonal frequency division of channels, or OFDM-modulation. At that, the carrier frequencies within each subchannel are modulated using QAM modulation, and PSK modulation is used for transmitting service information.

Several from N carrier frequencies of the OFDM signal are pilot signals which are used for frequency tuning and clock synchronization on the receipt side.

Modulation schemes with multiple carrier frequencies have the following features:

- Since there are many carriers, simple modulation schemes with a low modulation index M can be used to modulate them, and the same bit rate will be reached as with single-carrier modulation with high M value. Number of carrier frequencies can be tens or hundreds, but information symbol duration is much longer than in methods with single-carrier modulation. Low symbol rate allows using guard intervals (GI) between symbols to eliminate inter-symbol interference.
- Due to complex nature of SNR(f) dependency, SNR value will differ for different subchannels. Depending on SNR values in each subchannel, various carrier frequency modulation schemes can be applied.
- If there is narrow-band interference in the transmission band, for example, related to operation of other communication systems, part of the carrier frequencies may be completely disabled and not used for transmitting information. In this case, energy necessary for these carrier frequencies can be redistributed to carrier frequencies of other subchannels.

During transmission, all signal frequency components are added together to get instantaneous signal form. After receipt of the signal form, its instantaneous spectrum is restored. For these operations, mathematical algorithms are used for inverse fast Fourier transform (IFFT) and fast Fourier transform (FFT).

IFFT equation which is used to get a group signal in the time domain from the frequency domain with set of modulated signals $f_0, f_1, f_2, \ldots, f_{N-1}$ has the following form:

$$S(n) = \sum_{n=0}^{N-1} s(k) \sin\left(\frac{2\pi kn}{N}\right) - j\sum_{n=0}^{N-1} s(k) \cos\left(\frac{2\pi kn}{N}\right)$$

where k is frequency index among the set of N carrier frequencies; n is a certain moment of time; and $s(k)$ is spectrum value for k-th frequency.

For FFT, the reverse procedure is used. From some random radio signal $S(t)$, set of signals $s(k)$ with carrier frequencies $f_0, f_1, f_2, \ldots, f_{N-1}$ is formed. FFT equation has the following form:

$$s(k) = \sum_{n=0}^{N-1} S(n) \sin\left(\frac{2\pi kn}{N}\right) + j\sum_{n=0}^{N-1} S(n) \cos\left(\frac{2\pi kn}{N}\right)$$

where $S(n)$ are the frequency sine and cosine coefficients $2\pi k/N$.

To separate symbols from each other, a cyclic prefix is formed between them or, in other words, a guard interval (GI) lasting from 10% to 25% of the symbol's duration. The cyclic prefix is a copy of the "tail" of the symbol placed at its beginning. Figure 2.10 shows example of forming and transmitting OFDM symbols, and Fig. 2.11 illustrates principle of forming a cyclic prefix.

Let us consider block diagrams of OFDM modulator and demodulator shown in Figs. 2.12 and 2.13.

The schemes contain converter of serial bit sequence $b(t)$ to set of parallel sequences $b_0, b_1, b_2, \ldots, b_{N-1}$. Each sequence modulates a separate carrier frequency

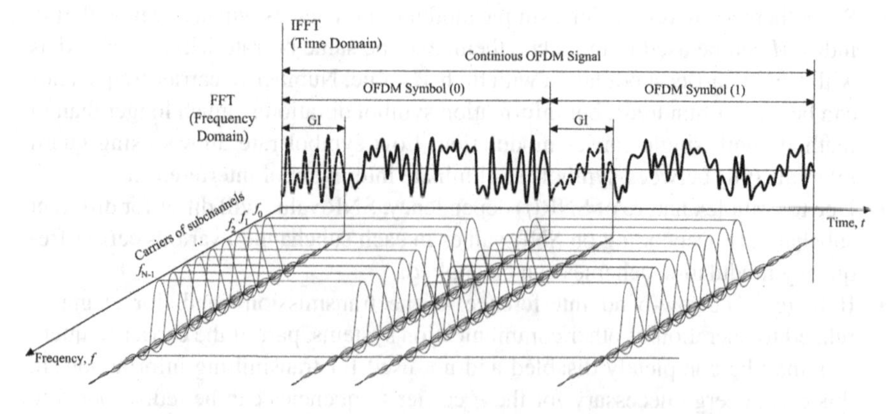

Fig. 2.10 Principle of forming and transmitting OFDM symbols (GI-guard interval)

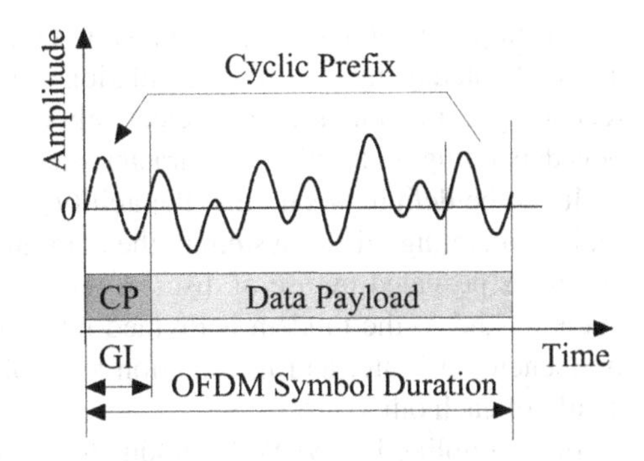

Fig. 2.11 Principle of forming a OFDM symbol cyclic prefix

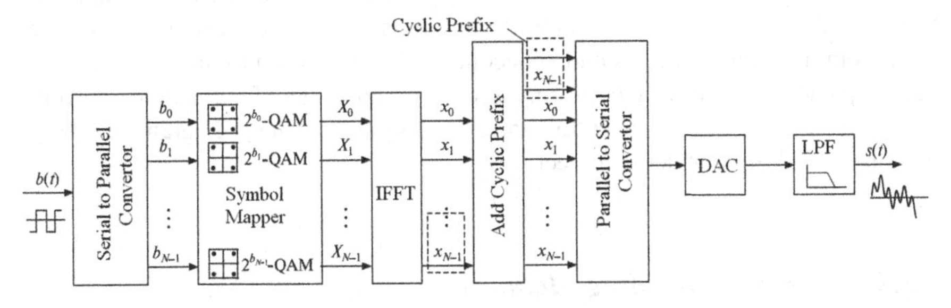

Fig. 2.12 Block diagram of OFDM modulator

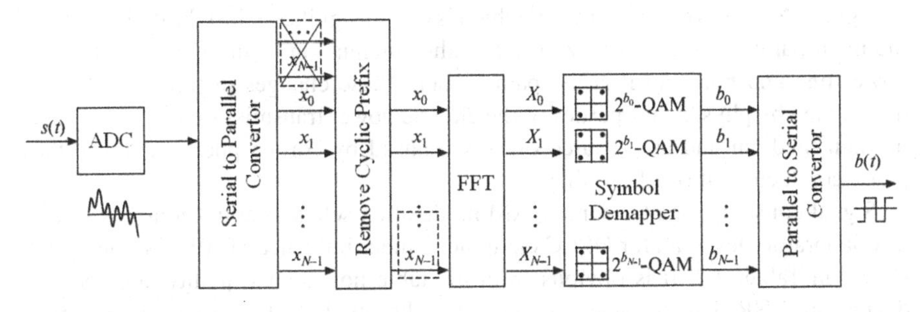

Fig. 2.13 Block diagram of OFDM demodulator

$f_0, f_1, f_2, \ldots, f_{N-1}$ in the symbol mapper unit, then the received signals pass through IFFT unit, and then a cyclic prefix is added to the symbol. Next, the signal components are combined into a single OFDM symbol, which is converted into a linear analog signal by DAC unit and a smoothing filter.

Reverse operation is performed on the receiver side. Linear analog signal together with noise is digitized in ADC unit, parallel signal is formed from the OFDM symbol, the cyclic prefix is excluded, the signals pass through FFT unit, and the signals of each of the subchannels $b_0, b_1, b_2, \ldots, b_{N-1}$ are demodulated. At output, initial bit sequence $b(t)$ is formed.

It is important to note that typically, in OFDM modulation all carrier frequencies are modulated using the same modulation scheme. This is a disadvantage of OFDM technology: it is impossible to selectively change modulation scheme for subchannels depending on frequency characteristic of the communication line.

It can be definitely said that the ability to change modulation scheme is a good tool for adapting DPLC system to the communication channel condition. This possibility is provided by use of discrete multi-tone DMT modulation, where the frequency band of the modem is divided into several subbands in which own carrier frequencies operate, and information symbols in each band are formed independently of each other.

In a simplified way, DMT-modulated modem can be represented as several OFDM modems running in parallel. This approach has the following advantages. First, DPLC system adjusts to non-uniform HF path frequency response and SNR of the communication line; second, it becomes possible to form frequency windows, i.e., frequency bands in which information is not transmitted, for example, to avoid mutual interference between DPLC and APLC systems. Block diagram of DMT-modulator will be considered in Sect. 2.8.

2.2.2 *Channel Encoding Methods*

As mentioned above, digital PLC channels are characterized by possibility of changing SNR in a very wide range, which is due to significant instability in time of the main parameters characterizing HF paths: attenuation of the HF path along a power line and noise level at the path output. These changes in time are mainly determined by physical properties of the electric power transmission line as a transmission medium and as a source of noise caused by corona phenomenon at the phase conductors of overhead line.

Signal constellations for some QAM modulation schemes at different SNR values (laboratory test data for DPLC equipment under influence of AWGN noise) are shown in Table. 2.2. It is obvious from the table how signal quality degrades at decreasing SNR that increases, respectively, the number of errors in the symbol recognition.

When operating a DPLC channel along overhead power line (OPL), broadband noise is defined by corona phenomenon at the overhead line phase conductors and has parameters which differ from AWGN parameters. At that, to achieve the same error rate under noise caused by corona effect, a larger SNR is required than in the case of AWGN noise. This issue was studied in [2, 19].

The higher the voltage class of the overhead power line, the more is the difference noise caused by corona and AWGN.

The essence of this phenomenon will be discussed in Chaps. 4 and 5. Here, for illustration, coefficient of bit error rate versus SNR for QAM modulation with different modulation indices is shown in Fig. 2.14. The dependences were obtained in [19] for simulation of DPLC channel under the influence of AWGN noise and

Table 2.2 Signal constellations for M-QAM modulation at various SNR values

SNR, dB	4QAM	16QAM	64QAM	128QAM
30				
25				Low SNR
20				Low SNR
15			Low SNR	Low SNR

corona noise, for 400 kV overhead power line. It can be seen from the graphs in Fig. 2.14 that, to obtain the same bit error rate when exposed to corona noise, a significantly higher SNR value is required than under influence of AWGN noise.

When complicating the communication system (by introducing channel encoding), the system performance can be significantly improved in a condition of high error rate. This is implemented by recovery of corrupted symbols using certain encoding algorithms with error correction. A simple signal constellation is then transformed into a code construction.

Let us mention the most known channel codes. One of the first correcting codes was Hamming code proposed in 1950. Convolutional codes, first created by L. M. Fink and V. I. Shlyapobersky, were proposed in 1955. Later, V. Hagelberger, A. Viterbi, J. Wozencraft, and A. E. Naifeh studied and development convolutional codes. Correcting cyclic Rid-Solomon code was developed in 1960 by I. Rid and G. Solomon. In 1966, D. Forni proposed to use turbo codes called "cascade codes,"

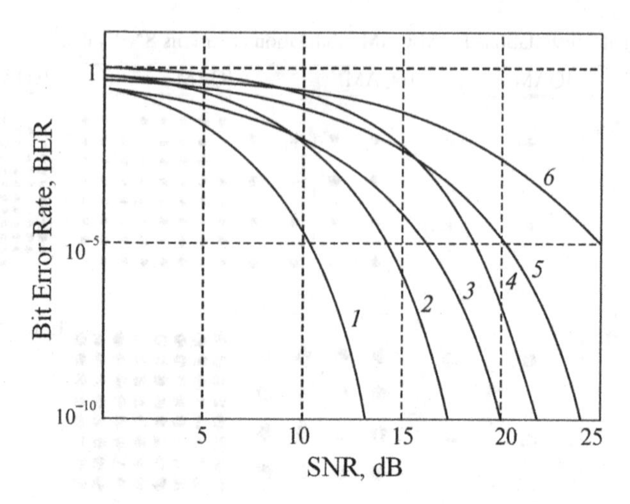

Fig. 2.14 SNR values required for different modulation schemes when operating DPLC in a channel with AWGN and corona noise: 1–4QAM (AWGN); 2–16QAM (AWGN); 3–4QAM (corona noise); 4–64QAM (AWGN); 5–16QAM (corona noise); 6–64QAM (corona noise)

which were implemented by means of sequential application of various convolutional and block codes.

LDPC codes (low-density parity-check code) are a class of codes with a low density of parity checks, proposed by R. Galagger in 1960 [8]. Trellis-coded modulation (TCM) proposed in 1982 by G. Ungerbock [29] is intended to increase Euclidean distance of PSK and QAM signal constellation points. Error correcting codes technologies are described in [6].

Digital transmission systems use the following channel encoding methods:

- Block encoding (FEC, forward error correction)
- Interleaving
- Convolutional encoding
- Scrambling

Block codes and interleaving encoding are used to increase noise immunity of digital transmission systems to pulse interference when group errors occur.

It should be noted that OFDM/DMT systems differ significantly from QAM systems: since the formation of OFDM/DMT symbol occurs in a narrow frequency band of each carrier, the symbol duration is tens or hundreds of times more than for QAM, and immunity to impulse noise is much higher, respectively (however, with more transmission delay).

At interleaving, symbols of one block are mixed according to a certain law with symbols of other blocks, so that long group errors are converted to single ones. At that, interleaving is possible both at the block level and at the level of symbols and bits in the same block.

Example of forming blocks with interleaving symbols is shown in Fig. 2.15 [25]. If some symbols in the original symbol block are damaged (4 out of 6), receiver will

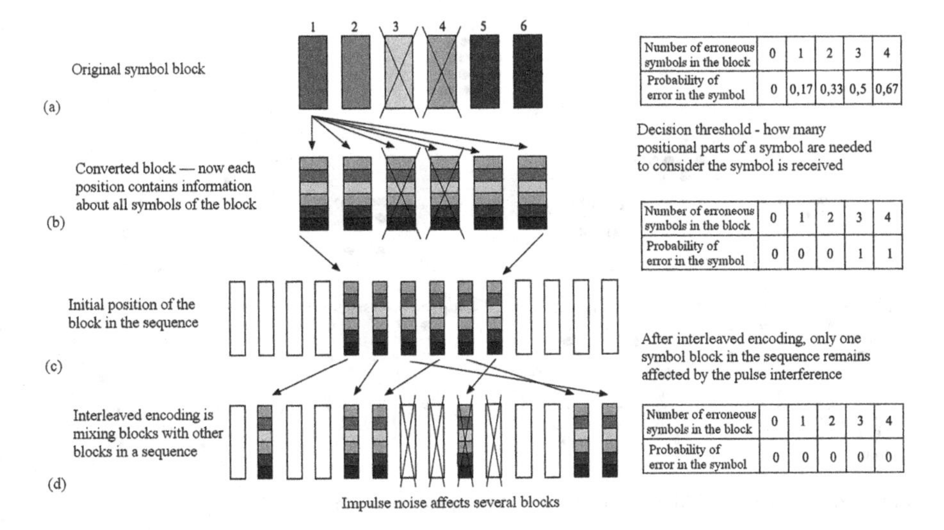

Number of erroneous symbols in the block	0	1	2	3	4
Probability of error in the symbol	0	0,17	0,33	0,5	0,67

Number of erroneous symbols in the block	0	1	2	3	4
Probability of error in the symbol	0	0	0	1	1

Number of erroneous symbols in the block	0	1	2	3	4
Probability of error in the symbol	0	0	0	0	0

Fig. 2.15 Principle of interleaved encoding

not be able to recover them (Fig. 2.15a). Converted block – now each position contains information about all symbols of the block – is shown in Fig. 2.15b. Number of positional parts of the block needed to regard that the block is received correctly corresponds to the decision threshold, e.g., 2 out of 3, 3 of 4, 3 of 5, 4 of 6, 4 of 7, etc. In our case, 2 out of 6 symbols are correct, and corrupted ones still cannot be restored.

Figure 2.15b shows position of affected block in the sequence of other blocks, and Fig. 2.15b shows principle of interleaved encoding: mixing symbols from one block with symbols from other blocks. It is made so that any 4-symbol impulse disturbance does not damage more than 2 symbols in the new sequence. And, accordingly, the block can be 100% recovered.

Use of interleaving significantly increases delay of transmitting information (in our case, 11/6... 14/6 times), but it allows eliminating group error of almost any length (in our case, 4 symbols). This is important for DPLC channels affected, for example, by lightning discharges. However, one bit or all bits of a symbol can be damaged. This means that interleaving restores a huge number of single, paired, and so on errors. In our case, the limit is 1–24 errors out of 36 bits of the block. And this is achieved without adding redundancy reducing the information rate similarly to other encoding methods.

Convolutional encoding is the main method for dealing with single errors. Principle of trellis encoding or TCM-modulation is based on use at the same time of PSK/QAM modulation and convolutional code. Moreover, sequence of changes in the transmitted signal amplitude and phase in each clock cycle depends on the input symbols values in the previous and subsequent clock cycles. The main idea is that each current symbol is transmitted with the maximum difference in Euclidean distance if compared with the previous and subsequent ones. In Fig. 2.16, scheme of

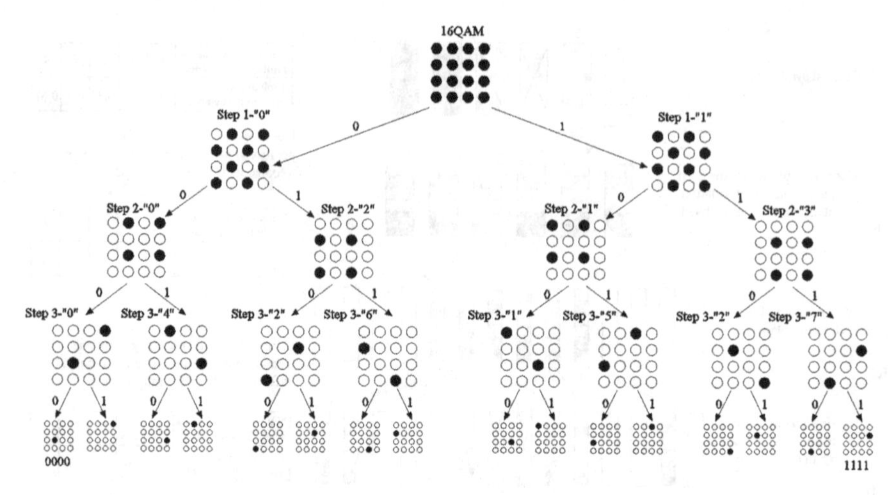

Fig. 2.16 Scheme of TCM encoding for 16QAM signal constellation

TCM encoding for points of 16QAM signal constellation is shown. Constellation points corresponding to code combinations which differ only by one bit are located close to each other since their values are encoded using gray code and single errors in neighboring code combinations can be corrected.

Graphs of symbol error rate versus SNR for modem of DPLC equipment which uses TCM modulation are shown in Fig. 2.17. As can be seen from the graphs, the use of TCM encoding allows achieving significant energy gain (the data are given for AWGN channel) [4].

Turbo codes and LDPC codes are promising for use in DPLC equipment. Practical implementation of these codes is consuming in terms of CPU computing resources, but it allows getting large error-correcting capability. Graphs of error coefficient versus SNR for various QAM modulation schemes, without encoding and with LDPC channel encoding, with encoding rate $(k, n) = 1/2$, are shown in Fig. 2.18 [23]. Encoding rate is characterized by parameters k and n. Where k is the number of information symbols received per clock cycle at the encoder input (user data), n is the number of symbols at the encoder output considering added check symbols.

Figure 2.19 shows graphs bit error rate versus SNR for 16QAM modulation with turbo encoding, with bit-interleaving BITCM (bit-interleaved turbo-coded modulation) in comparison with 4QAM modulation [20]. The graphs show characteristics of BITCM codes with different variations of generators for convolutional codes (n, k): (7, 5); (15, 13); (35, 23) [6]. The graphs are shown for the channel with AWGN and for corona noise on the 400 kV overhead power line. Similarly to LDPC codes, turbo codes allow getting significant energy gain.

In error correcting codes, symbols are used to form and process multidimensional data arrays, which are later used to generate output sequence. Redundancy makes it possible to define rules for transitions between symbols that increase the symbol Euclidean distance. DPLC receiver uses a decoder with Viterbi decoding

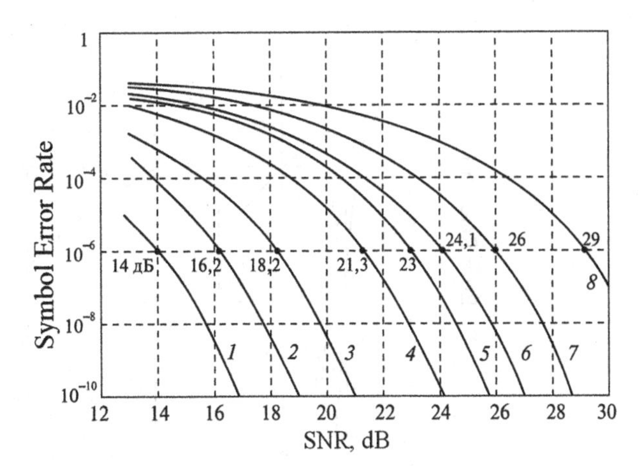

Fig. 2.17 SNR values required for various QAM and TCM modulation methods: *1* – 16TCM; *2* – 4QAM; *3* – 32TCM; *4* – 64TCM; *5* – 16QAM; *6* – 128TCM; *7* – 32QAM; *8* – 64QAM. (© 2020 IEEE. Reprinted, with permission, from Cesena and Castellani [4])

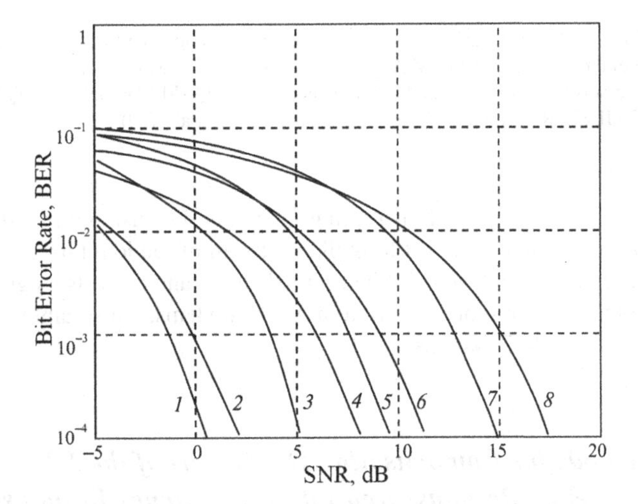

Fig. 2.18 Error probability vs SNR (for AWGN channel) for various modulation schemes without encoding and for channel encoding (LDPC): *1* – 8QAM-LDPC; *2* – 16QAM-LDPC; *3* – 32Q AM-LDPC; *4* – 64QAM-LDPC; *5* – 8QAM, uncoded; *6* – 16QAM, uncoded; *7* – 32QAM, uncoded; *8* – 64QAM, uncoded. (© 2020 IEEE. Reprinted, with permission, from Ortega-Ortega and Bravo-Torres [23])

algorithm based on the maximum likelihood principle [30], i.e., comparing the received code combination with a table of possible code combinations. To make a decision, receiver should process symbol sequences of a certain length, rather than single symbols or pairs. Hence, there is a delay, and the more it is, the more is the correcting ability of the code. Even if all symbols in the sequence are affected by single errors, Viterbi decoder selects the most appropriate combination of corrected output symbols in accordance with the known transition rules for the known Euclidean distance of the symbol sequence.

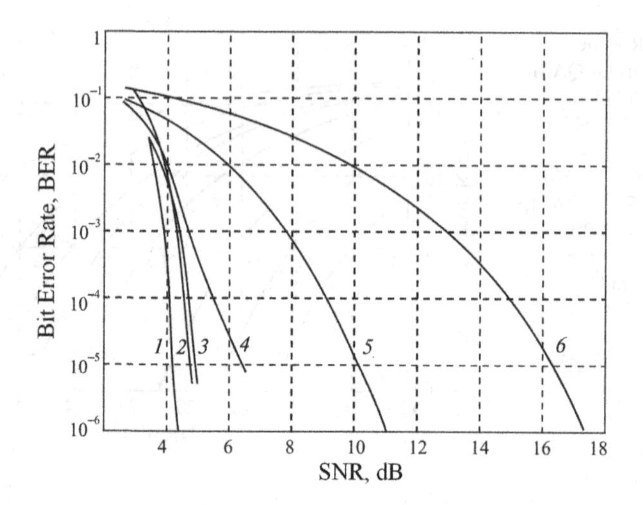

Fig. 2.19 BER vs SNR for various modulation schemes without encoding and for 16 QAM BITCM channel encoding: *1* – BITCM AWGN; *2* – (35, 23) BITCM (corona noise); *3* – (15, 13) BITCM (corona noise); *4* – (7, 5) BITCM (corona noise); *5* – 4QAM (AWGN); *6* – 4QAM (corona noise). (© 2020 IEEE. Reprinted, with permission, from Zajc et al. [20])

Scrambling is the process of encrypting incoming data using a pseudo-random encryption sequence that makes it look like a stream of random bits. The receiver, while knowing the cipher, decrypts the original signal and corrects single errors. It should be noted that scrambling is used in wired modems, but scrambling encryption is not used in DPLC systems.

2.2.3 *Methods for Compensating Distortions of the HF Path Frequency Response and Phase Frequency Response*

Transfer function of any four-pole device (including communication line, or, in the terminology adopted in PLCs, linear path) for sinusoidal signals is described by a complex value. Value of this number modulus describes signal level decrease at the end of the communication line relative to its beginning (attenuation), and angle value is the signal phase change between the same points.

Frequency dependence of attenuation is usually called frequency response (FR), and frequency dependence of the signal phase is called phase frequency response (PFR). Instead of PFR, group time delay (GTD) is often used. Non-distorting communication line (ideal transmission function) is a line (linear HF path) in which signal attenuation does not depend on frequency, and signal phase changes directly proportional to the signal frequency (GTD is the same for different frequencies). Transfer function of any real linear path is not ideal and depends on frequency in a complex way. This causes degradation of communication channel parameters.

For HF path of PLC channels, FR and PFR deviations from their average value (both increase and decrease) can be very large even within narrow frequency band. For example, in 4 kHz band under adverse conditions deviations can reach up to 10 dB/3 ms or more.

Imperfect transmission characteristics significantly impair signal reception due to changes in amplitude and phase (position) of the signal constellation points. This leads to errors in symbol recognition, especially for high-index modulation schemes. Appearance of non-uniformity depends not only on the communication channel characteristics, but also on parameters of the modem and power amplifier of the transmitting equipment. Echo signal of near- and far-end echo of own transmitter results in multiple copies of the signal at the receiver input. When DPLC system operates in the adjacent frequency mode, out-of-band emission, and non-uniformity of FR and PFR, reflected HF signals and noise from its own transmitter affect the signal reception.

The effect of FR and PFR distortion on error rate for PLC equipment modems was considered in [2]. Figure 2.20a shows 64QAM modulation signal constellation without distortion. To compare effect of low SNR and FR and PFR distortion, the same signal constellation but with low SNR value is also shown in Fig. 2.20b. It is evident that points of the constellation are randomly blurred: the smaller SNR, the lower density of points.

AFR non-uniformity leads to blurring constellation points (Fig. 2.20c): the more distance from the constellation center, the more non-uniformity.

When intermodulation distortions appear (e.g., when nonlinear distortions appear in the transmission path), points of the signal constellation look like concentric circles with approximately equal filling density.

In the case of non-uniform PFR characteristic, spherical symmetry of the constellation points relative to the constellation center is observed. A special case of the phase shift is IQ instability, which is a phase mismatch between in-phase and quadrature components of the signal [7]. So, a phase shift of just 1 degree for

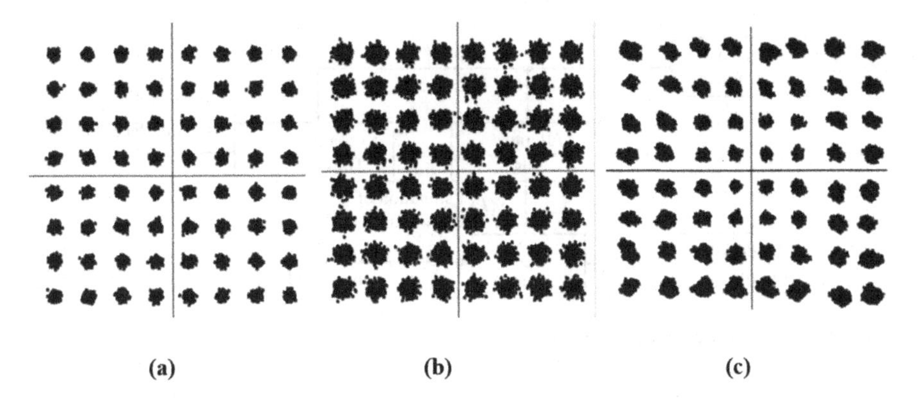

Fig. 2.20 Examples of signal constellation distortion: (**a**) normal signal constellation; (**b**) low SNR; (**c**) non-uniformity in FR

1024QAM scheme is equivalent to increasing required SNR value for bit error rate 10^{-6} by more than 8 dB.

To equalize the channel transmission characteristics, adaptive equalizers are used, and echo cancellers are used to suppress signal sent by own transmitter to own receiver. These devices are based on adaptive digital filters with a finite impulse response (FIR filters) which change their characteristics during operation in accordance with changes in characteristics of the communication channel.

Echo canceller scheme given in Fig. 2.21 shows how echo signal of own transmitter $x(k)$ is compensated by filtering through an adaptive filter. In the adder, the echo signal $+d(k)$ is added to the image of transmitted signal after filtering $-y(k)$. At the adder output, $a(k)$ signal is formed with the suppressed caused by the transmitter echo signal.

FR and PFR of the channel with equalizer should be independent of the frequency. For this, transfer characteristic of the equalizer should be reverse to the transfer characteristic of the communication channel. For the equalizer to be able calculate transmission characteristic of the communication line, a special "training" symbol sequence which is transmitted when establishing a connection between two modems is used. Equalizers can be without feedback, with feedback, or based on blind prediction. Training sequence is not required for operation of blind prediction equalizers.

Diagram of adaptive equalizer with feedback is shown in Fig. 2.22. To calculate weighting coefficients of the equalizer filter, adaptive filtering algorithm using the least mean square method (LMS, least mean square) is used [11]. The essence of the algorithm is as follows. The receiver initially knows what training sequence of symbols it should receive (training mode).

When a training sequence is received, standard deviation from the nominal symbol parameters is calculated. Thus, when deviation is known, it is possible to adjust the received symbol (estimation mode). When using a feedback equalizer, filter coefficients are updated considering correct symbols previously received at the

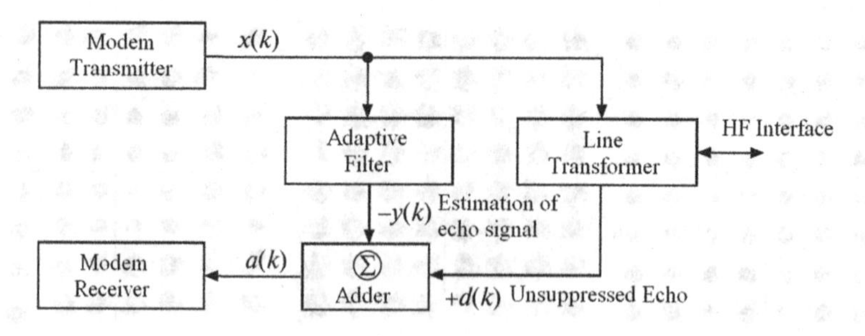

Fig. 2.21 Scheme of echo canceller

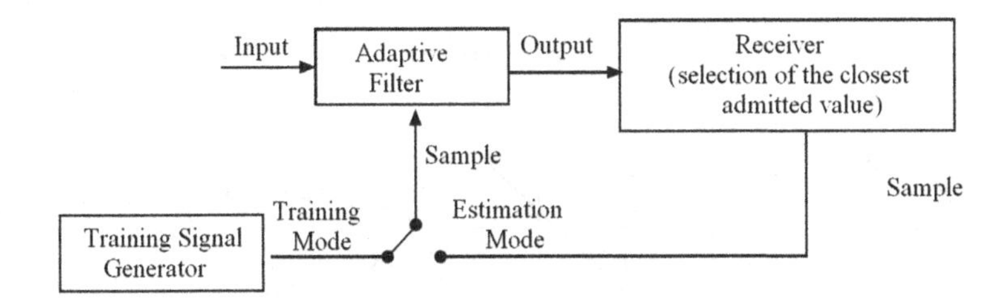

Fig. 2.22 Scheme of adaptive equalizer

input. This approach allows comparing and adjusting filter weights in the estimation mode. When power line is energized without any commutations in the electric grid, attenuation and noise level of the power line change smoothly in the time interval from tens of seconds to hours. To correct operation of the adaptive equalizer, a cyclic transmission of training sequence with a specified repetition period is used that can be called automatic equalizing of the channel characteristics.

2.3 DPLC Modems Bit Rate

Bit rate of information transfer by DPLC modems can be gross bit rate (bit rate on HF modem interface) and useful/ information bit rate (bit rate at data transfer interface). Gross bit rate is the sum of user data bit rate (useful rate) and bit rate required for transmitting service information. As a rule, useful bit rate of a simple modem is at least 90% of gross bit rate of the system. However, when using error correcting codes, gross bit rate remains unchanged, but useful bit rate decreases due to application of redundancy. When encoding rate is known, useful bit rate can be defined. For example, if encoding rate is 3/4, useful bit rate decreases by 25%.

Regardless of the modulation schemes, single-carrier or multi-carrier and code construction used, spectral efficiency of modems cannot be more than Shannon's limit. Maximum value of the spectral efficiency is defined as $n = \log_2(M)$.

Graphs of DPLC spectral efficiency for Shannon's limit and for modems of typical DPLC systems with BER 10^{-6} and 10^{-4} [13] are shown in Fig. 2.23. Note that these graphs are plotted for disturbances in the form of white noise without considering difference of white noise and corona noise.

Characteristics of code constructions used in the best DPLC modems are lower than Shannon's limit by only 1... 2 dB. This means that, at this stage of technical development, bit rate limit of HF modems has been reached. Using the graph in Fig. 2.23, it is not difficult to calculate approximately maximum theoretical bit rate for a given operating frequency band of the modem by simply multiplying W by η for a given values of SNR and BER.

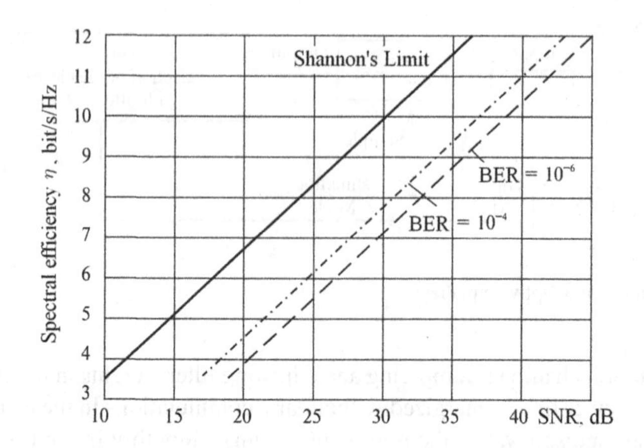

Fig. 2.23 Graphs of η(SNR) for typical DPLC modem. (IEC 62488-1 ed. 1.0 "Copyright© 2012 IEC Geneva, Switzerland. www.iec.ch" [All such extracts are copyright of IEC, Geneva, Switzerland. All rights reserved. Further information on the IEC is available from www.iec.ch. IEC has no responsibility for the placement and context in which the extracts and contents are reproduced by the author, nor is IEC in any way responsible for the other content or accuracy therein])

2.4 Dynamic Bit Rate Adaptation

The most important function of the DPLC equipment modem is its ability to dynamically adapt bit rate when SNR changes. At normal operation of digital system, bit error rate should not be more than 10^{-5}... 10^{-6}. If the bit rate is constant, reducing SNR leads to errors and, in the worst case, loss of clock synchronization between modems. Application of error correcting codes with interleaving ensures that information error rate at the level of 10^{-6} and synchronization between modems at BER up to 10^{-3} are maintained.

Under normal conditions or when a communication session is being establishing, two DPLC devices work in pairs, operating frequency band and bit rate of both devices are the same. One device is the main (Master) one; it controls operation modes of the opposite controlled device (Slave). Clock synchronization between modems is performed by means of service signals.

Modem bit rates at different steps of adaptation are initially defined by the modem settings or are set by user; they are a set of fixed values. To ensure modems operation with the required error rate when SNR decreases, it is necessary to switch it to a lower bit rate with another modulation scheme. Accordingly, when SNR increases and a certain period of monitoring its value expires, the modems should perform a reverse transition. This is the essence of principle of dynamic switching. How often transitions occur and time of modem operation at a reduced bit rate depends on duration of influencing factors, mainly weather. Number of adaptation steps depends on technical design of the modem and is usually not less than three.

The simpler code construction, the less modem synchronization time when switching between adaptation steps. For QAM/TCM modulation modems, synchronization time at change of bit rate is several hundred milliseconds. This has almost no effect on the process of user information transmitting. Therefore, for such modems, number of adaptation steps can reach ten or more. For example, modem can continue to work at adaptation step of 3... 4 dB when SNR changes by more than 40 dB, obviously, with data bit rate decrease or increase.

However, it is not practical to make the adaptation steps too small, for example, 1 dB, since errors occur during each transition when transmitting user data.

Modems which use multi-carrier modulation schemes and interleaving encoding can withstand SNR change up to 6... 10 dB without loss of synchronization and degradation of data transmission quality. However, delay in resynchronization for such modems when switching adaptation steps can reach few tens of seconds. Accordingly, user data transmission is interrupted at this time.

2.5 Time Delay in Transmitting Information

Delay of transmitting information caused by modem is an important factor which affects transmission of real-time signals, such as speech or operational control data. When transmitting speech, high delay impairs quality of its perception. When transmitting data signals with guaranteed delivery (e.g., TCP packets), amount of information which sender can transmit per unit of time decreases due to pauses for waiting receipt acknowledgments. Further, we will use for DPLC equipment the term "modem latency." Other sources of delay in transmitting information in DPLC channels will be discussed in Chap. 5.

Delay of HF signal transmission and reception in the DPLC equipment internal units, or modem latency, depends on the following main factors:

- Number of carrier frequencies used for transmission
- Use of interleaving in the error correcting schemes
- Algorithms for multiplexers and interface modules

Number of carrier frequencies determines width of filter required for each of them. The greater number of frequencies, the narrower the filter and the more time necessary to transmit a symbol. In this regard, single-carrier modulation schemes are better in comparison with multi-carrier ones. For example, in existing DPLC systems with OFDM modems, number of carrier frequencies can reach several hundred in the nominal transmission band up to 32 kHz. Let us suppose that operating frequency band of modem is 3.4 kHz. For single-carrier modulation, minimum symbol duration is 1/3. 4 $\approx$ 0.3 ms. For OFDM modem with 80 carrier frequencies, symbol duration is 80/3.4 $\approx$ 23.5 ms. But theoretically, the signal can "pass" 10 ms impulse disturbance which is not possible for QAM systems.

Relatively small additional delay is caused by linear convolutional encoding (correction of single errors) and data processing in the multiplexer and demultiplexer. Interleaving block encoding (second kind of interleaving) allows modems to handle error packets well. Previously formed symbols are grouped into blocks, replicas of the blocks are created, and the latter are mixed together. Decoder in the receiver performs reverse operations. As a result, a group error is transformed into single errors, which are corrected by subsequent decoders. Delay due to interleaving encoding depends on number of blocks to be processed and can reach tens of milliseconds.

Formula for approximate estimation of the latency in DPLC system t_{DPLC}, ms, can be expressed as follows:

$$t_{DPLC} = \frac{4N}{\Delta f} + t_{lin.code} + t_{interl.} + t_{bl.code} + t_{MDM}$$

where Δf is bandwidth of the carrier frequency filter, kHz; N is number of carrier frequencies; $t_{lin.code}$ is delay of the linear convolutional encoder-decoder, 10... 15 ms; $t_{interl.}$ is interleaving delay, 30... 50 ms; $t_{bl.code}$ is block encoder-decoder delay (forward error correction), 15... 25 ms; and t_{MDM} is delay of data processing information during multiplexing and demultiplexing, 10... 15 ms.

For example, for OFDM modem with an operating frequency band of 3.4 kHz and number of carrier frequencies $N = 80$, when using linear and block encoding with interleaving, minimum t_{DPLC} value will be 155 ms. For QAM modem with linear encoding, t_{DPLC} is only 21 ms. The above formula can be used for rough estimation of DPLC equipment latency; exact data depends on technical design of the modem.

2.6 Peak Factor of Modulated Signal

Peak factor (PAPR, peak to average power ratio) of modulated signal in relation to DPLC systems determines the difference between the levels of peak envelope power of transmitter and average signal power of HF signal.

The more modulation index, the more peak factor. Table 2.3 shows peak factors of symbols for various QAM modulation schemes. The least value has 4QAM or QPSK scheme – 0 dB; the upper limit of PAPR for QAM modulation $PAPR_{simb.QAM}$ is equal to 4.8 dB. Filters add to the peak-factor value of the modulated signal 3–5 dB due to such a phenomenon as filter transient process: appearance of spurious damping vibrations at the filter output, also called "ringing" of the filter.

As a rule, for single-carrier modulation methods, the signal peak factor is 5... 10 dB.

Table 2.3 *PAPR* values for M-QAM modulation

4QAM	16QAM	64QAM	256QAM	>1024QAM
0 dB	2.6 dB	3.7 dB	4.2 dB	4.8 dB

For multi-carrier modulation schemes, peak-factor depends on the filter bandwidth for each carrier frequency on transmission side. The greater number of carriers, the narrower the filters and the higher their impact on the peak-factor value. For multi-carrier modulated OFDM signal, peak factor is equal to

$$PAPR_{\text{OFDM}} = PAPR_{\text{symb.QAM}} + 10\log(N) + p_{\text{filter}}$$

where N is number of carrier frequencies; $PAPR_{\text{symb.QAM}}$ is peak factor of a symbol for specified modulation scheme used for carrier frequencies, dB; and p_{filter} is value of the filter "ringing."

For example, $PAPR_{\text{OFDM}}$, at 80 carrier frequencies, modulated by 16QAM, and at $p_{\text{filter}} = 5$ dB, will be 26.6 dB. To reduce peak factor, the following methods are used:

- By enabling trellis encoding (TCM) make that the carriers never have the same phase and amplitude of the resulting signal is as small as possible (component 10 log(N) is excluded from $PAPR_{\text{OFDM}}$ equation).
- Divide the carriers into separate blocks, while performing IFFT for each of them and combining the received signals into a linear signal.

As a result, it can be achieved that peak factor of the OFDM signal will be not more than 10... 15 dB. When using special codes, it is possible to achieve $PAPR_{\text{OFDM}} = 4... 6$ dB.

2.7 Comparison of Single-Carrier and Multi-carrier Modulation Methods Applied to DPLC Systems

Let us analyze advantages and disadvantages of different approaches to designing DPLC modems using single-carrier and multi-carrier modulations.

- Modulation parameters. QAM/TCM modems of DPLC equipment use one carrier frequency for each modem with bit number in symbol from 2 (4QAM) to 12 (4096QAM). In DMT/OFDM modems, number of carrier frequencies varies and can reach several hundred. For carrier frequency modulation, DMT/OFDM uses the same modulation schemes as QAM/TCM modems. Accordingly, in terms of technical implementation, DMT/OFDM modems are much more complex than QAM modems. However, to avoid affecting the signal constellation of the complex shape of the HF path frequency response and the group time delay of the HF path at high modulation indices, QAM/TCM modems require a complex adaptive equalizer.
- Peak factor of a modulated signal. Single-carrier modulation modems have a slightly lower PAPR than modems with multi-carrier modulation. This means that current level of transmitted modulated signal for HF interface is 3... 5 dB higher for QAM/TCM modems than for DMT/OFDM modems.

- Spectral efficiency. If SNR value is the same, spectral efficiency of both systems is the same. Systems with multi-carrier modulation have the best characteristics at high SNR value: more than 35 dB.
- Nominal reception and transmission bandwidth, adaptation of modems to non-uniformity of frequency response, and phase characteristics of the HF path. DPLC systems with DMT/OFDM modems are well adaptable to non-uniform frequency response of the HF path, since simple carrier modulation scheme with sufficiently large distance between constellation points does not impose any strict requirements for these parameters. There are no limitations of the channel bandwidth.
- For QAM/TCM modems, communication channel transmission characteristic should be equalized using adaptive equalizer; otherwise errors can occur in recognizing the signal constellation points. As a rule, QAM/TCM modems are used in DPLC channels frequency band of 4, 8, and 16 kHz. Wideband QAM/TCM modems for DPLC equipment with a nominal transmission bandwidth of more than 16 kHz are very rare in industrial implementation. Also DPLC systems which use 2–4 separate QAM/TCM modems with a nominal frequency band of 4... 16 kHz each, running in parallel, exist. This makes such systems similar to the simplest DMT/OFDM modems.
- Bit rate adaptation. Due to simple design, QAM/TCM modems can have many adaptation steps, more than 10, and can switch from one bit rate to another in a few tens to hundreds of milliseconds. Change in bit rate occurs when SNR is reduced by 3... 4 dB.
- For DMT/OFDM systems, number of adaptation steps is usually less and bit rate changes when SNR is reduced by 6... 10 dB. In the absence of interleaving block encoding, the rate switching time is comparable to QAM/TCM modems, and the switching time increases significantly when using interleaving.
- DPLC equipment latency. The greater number of carrier frequencies and the more complex is used channel encoding methods, the more modem's latency. For QAM/TCM systems, the delay is 20... 100 ms regardless of the nominal transmission bandwidth. For DMT/OFDM systems, the delay depends on number of carrier frequencies, nominal transmission bandwidth, and application of interleaving encoding. At nominal transmission bandwidth of 12 kHz or more, DMT/OFDM system delay is comparable to the delay of QAM/TCM systems. In the 4 kHz frequency band, the delay can reach 160... 200 ms, in the 8 kHz band – 80... 100 ms, respectively [16].
- Noise immunity of the system under white noise disturbance. In the case of white noise, both systems have the same error probability.
- Noise immunity of systems under impulse disturbances. We have explained above that QAM/TCM system has a low immunity to impulse disturbances. It leads to burst errors and even loss of synchronization between modems.
- DTM/OFDM system immunity to impulse disturbances will depend on presence of forward error correction function which can improve the system's immunity to bursts lasting up to hundred milliseconds, for example, to those caused by linear switch operation. In case of lightning discharges or operation of overhead

line bus disconnectors, duration of impulses can be from several hundred milliseconds to several seconds. In this case, DTM/OFDM system also cannot continue steady operation.

- Resistance to narrowband interference. Often, several narrowband signals can be detected in DPLC channels with a level significantly higher than noise level in the line. These are rare, but still occurring cases of PLC channels operation.
- Narrow-band interference negatively affects operation of modems with single-carrier modulation schemes. When additive interference is applied, "holes" or nonexistent points appear in QAM modulation signal constellation which cannot be recognized by the modem.
- In DMT/OFDM modem signals, narrowband interferences damage only one or few of many carrier frequencies. If there is a function for forming frequency windows or notches, these carriers are ignored or disabled and data transmission continues serviceable carriers with a slight decrease in total bit rate.

It is difficult to answer unequivocally to the question which DPLC systems, with single-carrier or multi-carrier modulation, are better. We can answer this question as follows: systems with single-carrier modulation methods should be given priority when using channels with an operating frequency band 4–8 kHz. Systems with multi-carrier modulation methods are better in channels with a bandwidth of 8–12 kHz or more.

2.8 Wideband Modems Based on DMT Modulation

The idea of using DMT modulation in DPLC equipment [1, 14] appeared long time ago [24], but did not get practical use. Interest in use of DSL modems in DPLC systems has resumed relatively recently with development of wideband digital PLC systems with packet switching (W-DPLC equipment). Comprehensive studies of the technology continue [17, 18].

Generalized scheme of DMT modulator is shown in Fig. 2.24. Incoming data sequence is represented as N parallel sequences each of which is encoded by a block code, such as Rid–Solomon one. Then, if necessary, interleaving is performed. Then, using convolutional (trellis) encoding, peak-factor is reduced and energy distributed to the carrier frequencies is calculated. Next, the scheme uses adaptive load redistribution algorithm for carrier frequencies which ensures the same transmission stability in each subchannel. After that, information is finally distributed to the carrier frequencies and sent to OFDM modulators (separately for each band) that forms a group linear signal.

The system dynamically controls modulation schemes, and energy distributed to the carrier frequencies enables and disables the carrier frequencies with forming frequency windows.

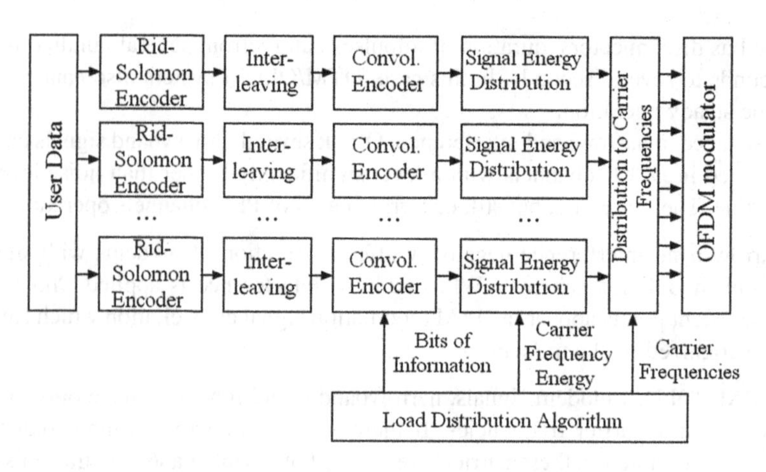

Fig. 2.24 DMT modulator

Use of DMT-modulation allows implementing modems which can theoretically occupy entire PLC frequency spectrum and achieving megabit data bit rates. The question is in the availability of frequencies. In addition to forming notches, DMT modulation allows asymmetric operating frequency bands for both transmission and reception directions. This is promising from the point of view that the main information flow is transmitted from substation to dispatch center, and only requests for information or acknowledgments of packets are transmitted from the dispatch center to the substation.

Figure 2.25 shows an example of two DMT signals location in the PLC frequency range for two transmission directions. Here, notches, transmission band asymmetry, and different spectral efficiency in individual frequency bands can be clearly seen.

Gross bit rate depends on SNR and is calculated separately for each carrier frequency. It should be noted that, for wideband PLC systems, bandwidth of the receiver filter is equal to the width of entire frequency range in which the equipment can operate. Signal reception is based on use of high-performance parallel analog-to-digital converters. Use of digital filters with a bandwidth of several hundred hertz for each carrier frequency allows for effective filtering of noise.

Number of carrier frequencies in each subchannel depends on the specific technical design of the modem. Naturally, the wider operating frequency band and the greater number of carrier frequencies, the lower the HF signal output level at the same transmitter power. Approximately, transmission level $p_{\text{TX(DMT)}}$, (dBm) of modulated signal in HF interface can be calculated using the following formula:

$$p_{\text{TX(DMT)}} = 10\log\left(\frac{P_{\text{PEP}} - \sum P_{\text{Sys}}}{P_0 N}\right) - \text{PAPR} \tag{2.1}$$

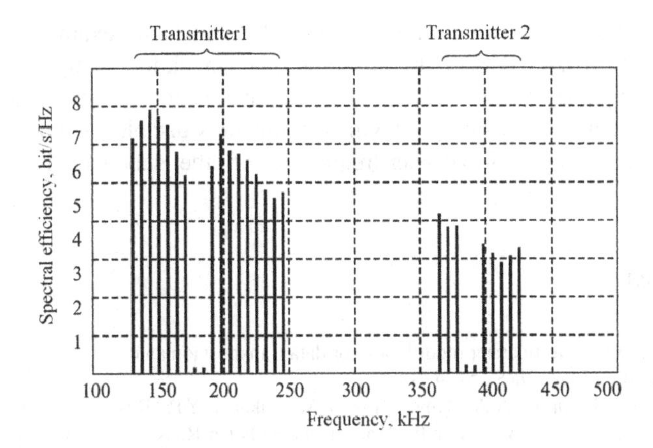

Fig. 2.25 Example of subchannel distribution in DMT modulation systems

where P_{PEP} is peak envelope power of the transmitter, W; N is number of carrier frequencies; P_0 is a zero-power level of 1 mW; P_{Sys} is the sum of powers of all service signals such as pilot signals; and PAPR is the peak factor of the modulated signal.

Let us analyze an example. Assume the bandwidth of the digital filter for each carrier frequency is 200 Hz, operating frequency band is 64 kHz, the system uses two pilot signals, and each occupies a band of 1 kHz. Peak envelope power of the transmitter is 100 W, total power of the pilot signals is 20 W, and the modulated signal peak -factor is 10 dB; 310 carrier frequencies can be contained in 62 kHz band. Substituting the data in (2.1), we obtain that the HF signal output level in HF interface will be 14.1 dBm in the entire operating frequency band of the modem.

The transmission bit rate will be as follows:

$$B_{\mathrm{DPLC(DMT)}} = \sum_{i=1}^{N} \Delta f \eta_i$$

where Δf is carrier's frequency band (of the modem), Hz and η_i is spectral efficiency of each carrier frequency.

Spectral efficiency depends on code construction used for carrier frequency modulation. Digital PLC systems with packet switching being now only example of possible use of DMT modems have very interesting difference from classic DPLC systems in terms of adapting transmission bit rate in the case of SNR change.

Modem of classic DPLC equipment has a fixed number of adaptation steps. In classic systems, transmission bit rate for each step is set to a certain value, and as the error rate increases, the system tries to keep running at the specified rate until number of errors exceeds a certain set limit. In DMT modems, each subchannel adapts the carrier frequency modulation scheme independently of the other subchannels. The process is fully automated, and the system independently selects which modulation scheme to use for different subchannels, for example, QAM 4, 8, 16, 64, 256, 512, 1024, 2048, 4096. If in some subchannel, for example, in the upper part of the frequency band, SNR value is not enough to work with the simplest modulation scheme, this subchannel will be completely disabled and the transmitter power will be redistributed to carrier frequencies of other subchannels.

References

1. Bingham, J.A.C.: Multicarrier modulation for data transmission an idea whose time has come. IEEE Commun. Mag. **28**(5), 5–14 (1990)
2. Braude, L.I., Philippov, A.A., Kharlamov, V.A., Shkarin, Y.P.: Research of discrete information transmission along power line carrier channels (in Russian). Electr. J. (8), Moscow, 8–14 (1999)
3. Cahn, C.: Combined digital phase and amplitude modulation communication. IRE Trans. Comput. Syst. **CS-10**, 90–95 (1962)
4. Cesena, F., Castellani, A.: A novel system for 64 kbit/s digital transmission over high voltage power lines. Paper presented at the 2003 IEEE Bologna Power Tech Conference Proceedings, Bologna, Italy, 23–26 June 2003, vol. 3, pp. 6. https://doi.org/10.1109/PTC.2003.1304494
5. Chang, R.W.: Synthesis of band-limited orthogonal signals for multichannel data transmission. Bell Syst. Tech. J. **46**, 1795–1796 (1966)
6. Clark Jr., G.C., Cain, J.B.: Error-Correction Coding for Digital Communications. Springer US, Springer (1981)
7. Distributing High-Definition TV on Coax. Blonger Tongues Lab. Inc. InfoComm https://www.blondertongue.com/page/media/InfoComm-2010.pdf (2010). Accessed 01 June 2020
8. Gallager, R.G.: Low density parity check codes. Sc.D. thesis, MIT, Cambridge (1960)
9. Hanzo, L., Ng, S.X., Keller, T., Webb, W.: Quadrature Amplitude Modulation. From Basics to Adaptive Trellis-Coded, Turbo-Equalized and Space-Time Coded OFDM, CDMA and MC-CDMA Systems, John Wiley & Sons Ltd, The Atrium, Southern Gate, Chichester West Sussex P019 8SQ, England(2004)
10. Hartley, R.V.L.: The transmission of information. Bell Syst. Tech. J. **3**, 535–564 (1928)
11. Haykin, S., Widrow, B.: Least-Mean-Square Adaptive Filters. Wiley, New York (2003)
12. Hirosaki, B.: An orthogonally multiplexed QAM system using the discrete Fourier transform. IEEE Trans. Commun. **COM 29**, 983–989 (1981)
13. IEC 62488-1 Power line communication systems for power utility applications – Part I. Planning of analogue and digital power line carrier systems operating over EHV/HV/MV electricity grids. International Standard, IEC (2011)
14. Kalet, I.: The multitione channel. IEEE Trans. Commun. **37**, 119–124 (1989)
15. Kotelnikov, V.A.: On the transmission capacity of "ether" and wire in electro communications, Izd. Red. Upr. Svyazzi RKKA (1933), Reprint in Modern Sampling Theory: Mathematics and Applications, Benedetto, J.J., Ferreira, P.J.S.G. (eds.), Birkhauser, Boston (2000)
16. Merkulov, A.G.: Technologies of packet networks organization via high voltage digital power line carrier channels (in Russian). Hot Line Telecom, Moscow (2017)
17. Merkulov, A.G., Adelseck, R., Buerger J.: Wideband digital power line carrier with packet switching for high voltage digital substations. Paper presented at the IEEE International Symposium on Power Line Communications, Manchester, UK, 8–11 April 2018
18. Merkulov A.G., Kussyk, J. Frankenberg, R.: Distinctive features, characteristics and field tests of the WDPLC systems. Paper presented at the IEEE Global Power, Energy and Communication Conference, Cappadocia, Turkey, June 12–15 2019

19. Mujčić A., Suljanović, N., Zajc, M., Tasič, J.F.: Error probability of MQAM signals in PLC channel. Paper presented at the 11th International Electrotechnical and Computer Science Conference ERK 2003, Portorož, Slovenia, 25–26 September 2003
20. Zajc, M., Mujčić, A., Suljanović, N., Tasič J.: High voltage power line constraints for high speed communications. Paper presented at the 12th IEEE Mediterranean Electrotechnical Conference (IEEE Cat. No.04CH37521), Dubrovnik, Croatia, May 12–15, 2004, vol. 1, pp. 285–288. https://doi.org/10.1109/MELCON.2004.1346840
21. Murphy, E., Slattery, C.: All about direct digital synthesis, ask the application engineer. Analog Dialogue. **38-08**, (2004)
22. Nyquist, H.: Certain factors affecting telegraph speed. Bell Syst. Tech. J. **3**, 324 (1924)
23. Ortega-Ortega, A.L, Bravo-Torres, J.F: Combining LDPC codes, M-QAM modulations, and IFDMA multiple-access to achieve 5G requirements. Paper presented at the 2017 International Conference on Electronics, Communications and Computers (CONIELECOMP), Cholula, 2017, pp. 1–5. https://doi.org/10.1109/CONIELECOMP.2017.7891828
24. Pighi, R., Raheli, R.: On multicarrier signal transmission for high-voltage power lines. Paper presented at the IEEE International Symposium on Power Line Communications, Vancouver, Canada, 6–8 April 2005
25. Romanov, S.E.: Design of DPLC channels and networks (in Russian). http://romvchvlcomm.blogspot.com/2007/08/blog-post_8270.html (2005). Accessed 01 June 2020
26. Proakis, J.G.: Digital Communications, 5th edn. Avenue of the Americas, New York, NY 10020 McGraw-Hill (2007)
27. Shannon, C.E.: Communication in the presence of noise. Proc. Inst. Radio Eng. **37**(1), 10–21 (1949)
28. Sklar, B., Ray, P.K.: Digital Communications. Fundamentals and Applications, 2nd edn. Pearson, New York (2014)
29. Ungerboeck, G.: Channel coding with multilevel/phase signals. IEEE Trans. Inf. Theory. **IT-28**, 55–67 (1982)
30. Viterbi, A.: Error bounds for convolutional codes and an asymptotically optimum decoding algorithm. IEEE Trans Inf Theory. **13**(2), 260–269 (1967)
31. Weinstein, S.B., Ebert, P.M.: Data transmission by frequency division multiplexing using the discrete Fourier transform. IEEE Trans. Commun. Technol. **19**, 628–634 (1971)

Chapter 3
Multiplexers and Network Elements of DPLC Equipment

3.1 Multiplexers

DPLC equipment includes a multiplexer with time division of signals and/or a network element. The multiplexer can be made as FPGA located on the CPU module board or as a separate module in the system chassis.

Basic functions of a multiplexer are as follows:

- Multiplexing incoming signals received from user voice frequency interfaces and data interfaces into a single aggregated digital signal transmitted to the modem
- Demultiplexing digital signal received from the modem to transmit information to the user interfaces
- Prioritization of user interface signals
- Routing of user signals in transit connections at the aggregated signal level.

3.1.1 Typical Schemes for Multiplexer Application in DPLC Equipment

Let us consider typical schemes of using a multiplexer as a part of DPLC equipment. Figure 3.1 shows diagram using multiplexer built into the CPU module. In this case, the multiplexer is responsible for processing data signals received from the user interfaces, but it does not process speech signals. This scheme is widely used in DPLC equipment when part of the operating band is used for analog speech transmission, and the remaining part is used for the digital modem operation and data transmission. Multiplexer can use various data interfaces RS-232 and X.21, and it processes data coming from the network element. Various data collecting devices are connected to DPLC equipment via RS-232 interfaces, for example, data

A. G. Merkulov et al., *High Voltage Digital Power Line Carrier Channels*,
https://doi.org/10.1007/978-3-030-58365-1_3

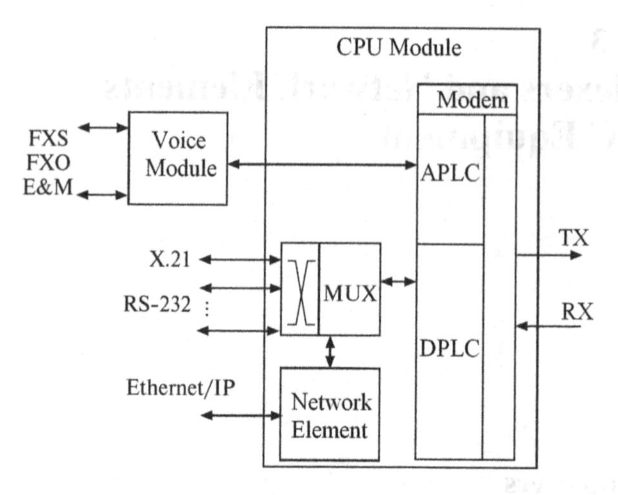

Fig. 3.1 Multiplexer built-in in CPU module

from telecontrol system can be transmitted via IEC 60870-5-101 protocol. X.21 interface is used to connect various external multiplexers and IP routers. Detailed information about parameters of data interfaces can be found in [1].

Figure 3.2 shows diagram using multiplexer as a separate module. Speech signals are sent from voice modules to the multiplexer which compresses them using low bit rate speech codecs.

In some industrial models of DPLC equipment, multiplexers have a framed two-megabit E1 interface for connecting PABX or external PDH multiplexers.

In this case, multiplexer of DPLC equipment compresses digital voice signal transmitted from the external device in 64 kbps timeslots of the E1 stream.

A transit interface which is used for transiting speech and data signals at the level of aggregated digital signal, rather than voice frequency interfaces or user data interfaces can also be implemented in the multiplexer.

Network elements are used for connecting various network devices to DPLC equipment via Ethernet interface. Principles of network element operation will be discussed in the next section. As a rule, most DPLC systems use transparent Ethernet bridge which does not perform any operations with Ethernet frames or IP packets at the protocol level.

Data bit rate on the Ethernet interface side is 10 or 100 Mbps. The multiplexer determines the rate of data transmission from a network element to the communication channel considering processing of other data signals and speech signals.

The number of data and speech signals processed by the multiplexers depends on design of multiplexers of DPLC equipment.

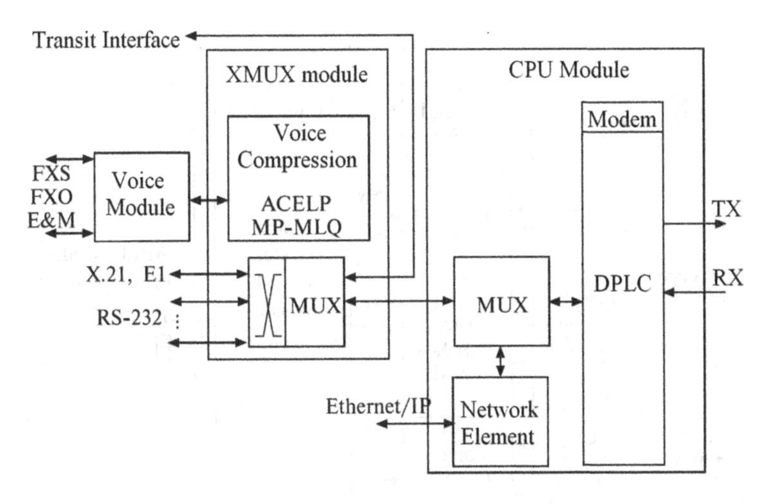

Fig. 3.2 Multiplexer designed as a separate module

3.1.2 *Processing of Speech Signals. Low Bit Rate Speech Codecs*

Figure 3.3 shows block diagram of the voice module which includes several voice ports:

- Two-wire FXS subscriber terminal to which analog telephones are connected.
- Two-wire FXO station terminal to which telephone station line is connected.
- Four-wire connection line with E&M signaling. It is used for connecting to PABX via E&M interface.

Depending on the voice module design, types of voice frequency interfaces either can be selected programmatically or can be hard-linked to port numbers.

Analog voice signal passes through analog-to-digital converter and G. 711 pulse code modulation speech codec and is converted to digital signal at bit rate of 64 kbps.

This signal is transmitted via the equipment internal bus either to CPU module for conversion and transmission in analog APLC mode or to multiplexer module for further conversion using various low bit rate speech codecs typically G. 729 [2] or G. 723. 1 [3].

For demultiplexing of speech signals, the multiplexer performs inverse transformation.

Most low bit rate speech codes are based on two algorithms: ACELP, algebraic code-excited linear prediction, and MP-MLQ, multi-pulse maximum likelihood quantization.

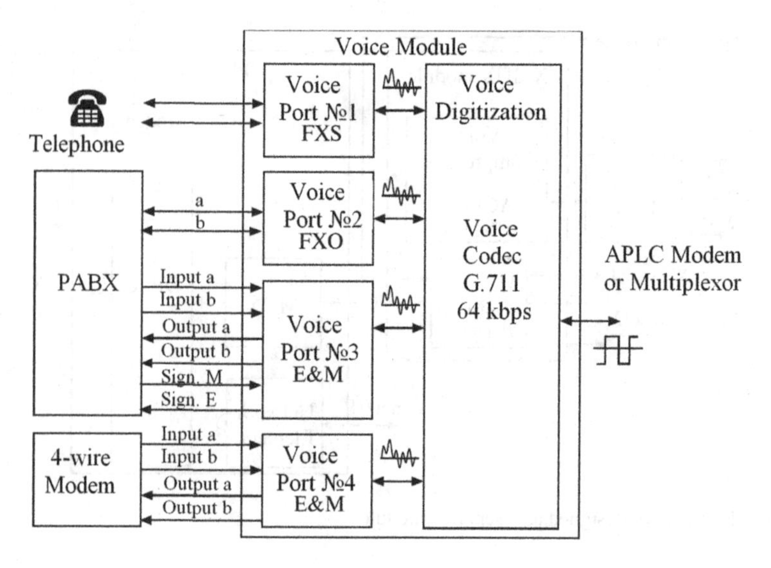

Fig. 3.3 Block diagram of the voice module

Using G. 729 codec (ACELP) makes it possible to convert original speech signal at 64 kbps to a 6.4 or 8 kbps signal.

When using the G. 723.1 codec (MP-MLQ), voice signal bit rate is 6.3 kbps, and for G. 723.1 (ACELP), the value is 5.3 kbps only.

It is important to note that these values relate directly to the codec characteristics. Required bit rate of speech signal transmission will be higher, as a rule, by no more than 10%, due to transmitted service information of multiplexers.

3.1.3 Prioritization of Information Signals

Let us analyze how the multiplexer prioritizes signals coming from user interfaces. Under normal conditions, the modem operates at a set bit rate, and this bit rate is enough for simultaneous transmission of a certain number of speech and data signals. When SNR is reduced, modem reduces transmission bit rate, so simultaneous transmission of this number of signals is impossible. Transmission of low-priority signals should be blocked. Or low-priority signals can be transmitted if the channel is not busy with high-priority signals, for example, if one of the voice channels is not busy. Channel priorities are assigned programmatically when configuring DPLC device.

Figure 3.4 shows an example of graph of bit rate versus SNR indicating modem adaptation steps and transmitted signals [4]. Speech and data signal bit rates are shown conditionally. In total, in this example, three speech signals are transmitted at a rate of 7.2 kbps, a data signal from low-bit rate RS-232 interface at a rate of

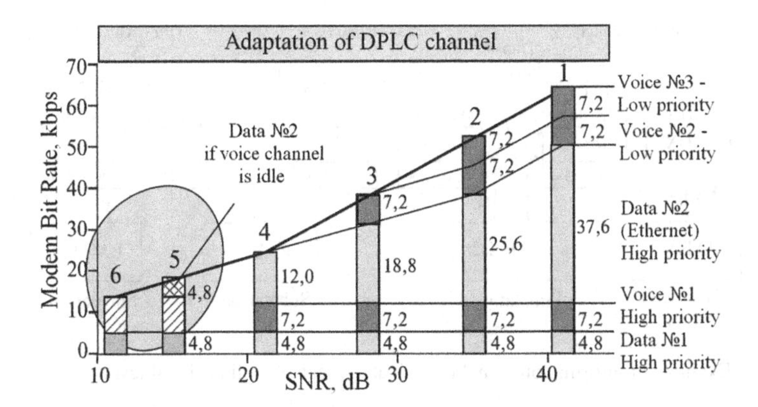

Fig. 3.4 Steps of modems bit rate adaptation

4.8 kbps and a data signal from Ethernet interface of the network element which accounts for the entire remaining bandwidth of the communication channel. At normal conditions, the value is 37.6 kbps.

The primary bit rate of DPLC equipment modem is 64 kbps, the second is 52 kbps, the third is 38 kbps, the fourth is 24 kbps, the fifth is 16.8 kbps, and the sixth is 12 kbps.

The figure clearly shows that, with a decrease in SNR and decrease in the modem bit rate, multiplexer first reduces bit rate of data transmission from the network element, then turns off non-priority voice channel 3, then turns off non-priority voice channel 2. This also reduces the bit rate of data transfer from the network element. As a result, there is only one data signal of the highest priority, voice signal № 1, and data from the network element (4th adaptation step). At the 5th adaptation step, data from the network element can only be transmitted when the voice channel is not busy, at the 6th step, Ethernet data transmission is blocked. If SNR further decreases, synchronization between DPLC equipment modems is lost, and user information transfer is interrupted.

When initial SNR value is restored, modems increase the bit rate, and multiplexer restores signal transmission according to their priorities.

3.1.4 Transit of Information Signals

Transit of information signals in intermediate nodes using classical multiplexers with time division of signals, as a rule, is performed through user data interfaces and voice frequency interfaces.

In other words, in transit points, all signals are demultiplexed and transmitted via wire connections.

A big disadvantage of this approach is the need to perform compression and decompression of speech signals several times that significantly reduces speech

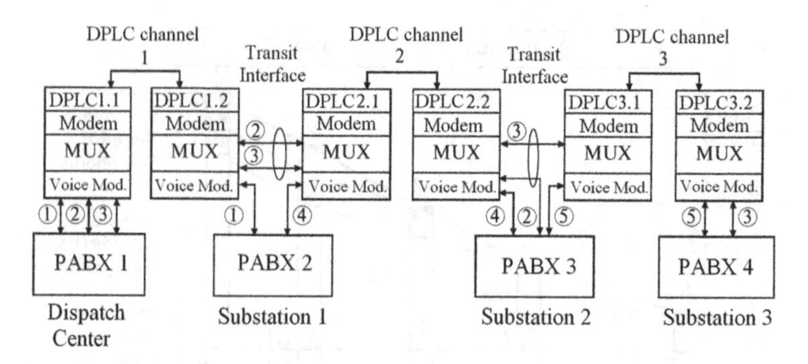

Fig. 3.5 Example of information signal routing at aggregated digital signal level

perception quality. It should be also kept in mind that wire transition requires the presence of voice modules and free data interfaces in all devices.

Scheme of several DPLC systems interaction in transit nodes when transmitting speech signals is shown in Fig. 3.5. Voice signals are transmitted between a dispatch center and three substations. Voice channels 1, 4, and 5 connect neighboring PABX by connecting to them via the VF interfaces. Voice signals 2 and 3 are transited from the PABX in the dispatch center to remote substations.

In the example, multiplexers of DPLC systems are interconnected through a transit interface allowing connecting them in a single bus, for example, via RS-422 interface which can provide a short distance bit rate up to 10 Mbps. When the modem bit rate is tens or even hundreds of kbps, 10 Mbps speed will be enough to transfer information between devices at the aggregated digital signal, without the voice and data signals termination on the user interfaces. Information is transmitted between devices according to the multiplexer routing table.

In addition to the above scheme, there is an option of scheme for transit of analog signal of DPLC modem without converting it into aggregated data signal. In this case, modems are installed only at the edges of the DPLC channel. This approach minimizes delay for transit connections due to the absence of modem delays. But it has disadvantages, such as summation of different HF path noise, possible accumulation of AFR /PFR non-uniformity, adding channels at intermediate substations is possible only by expanding the nominal frequency band of the PLC channel.

3.2 Network Elements

In Ethernet/IP networks, information frames are formed with adding of service information from the physical, data link, network, and transport layer protocols. When transmitting guaranteed delivery data, IP and TCP (Transport Control Protocol) protocols are used: data from telecontrol systems (TC), power metering systems (PM) systems, e-mail, Internet, and file exchange. When transmitting non-guaranteed delivery data, IP and UDP (User Datagram Protocol) protocols are used – various equipment remote monitoring data using SNMP Protocol. When transmitting voice signals using VoIP (Voice over IP) technology, information frames contain IP, UDP, and RTP (real transport protocol) protocol headers. On the data link layer Ethernet protocol or Ethernet frames encapsulation by PPP (Point to Point), Frame Relay, or HDLC (High-Level Data Link Control) protocol are used. To learn more about the information transmission principles for IP networks, we recommend referring to [1].

Most modern DPLC systems can transmit packet traffic from Ethernet networks. Interaction of DPLC equipment and network devices is performed by means of network elements integrated into DPLC equipment. Let us analyze structure of network elements and technologies used in them.

3.2.1 Structure of DPLC Equipment Network Element

Network elements of DPLC systems are based on technologies and protocols used in any other network devices: hubs, switches, and routers. An example of network element block diagram is shown in Fig. 3.6.

LAN (Local Area Network) ports are used to connect external network devices via Ethernet 10/100 Base-T electric interfaces or via optical interfaces, such as 100 Base-FX.

Packet processing is performed by network traffic processing controller or IP controller. Interface for DPLC system interaction with multiplexer or modem is usually a part of common internal bus for exchanging DPLC device information. In addition to the user LAN interfaces, LCT (Local Craft Terminal) LAN interface is in the network element for connecting to central processor of the DPLC equipment via TCP/IP connection. In Fig. 3.7, functional diagram of DPLC equipment interaction with router is shown.

Like any network equipment, the DPLC system network element operates using various Open Systems Interconnection Basis Reference Model (OSI) protocols. In

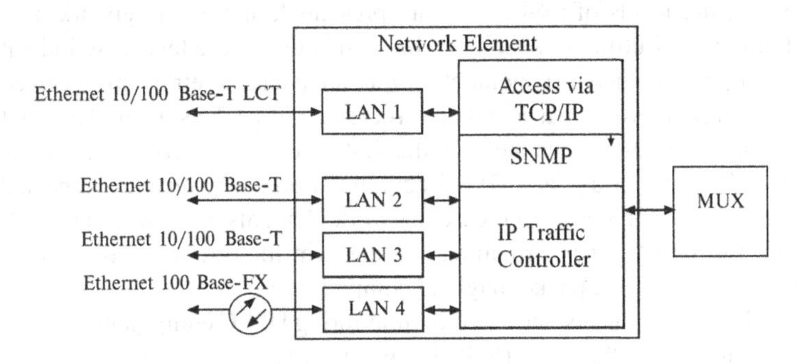

Fig. 3.6 Block diagram of DPLC equipment network element

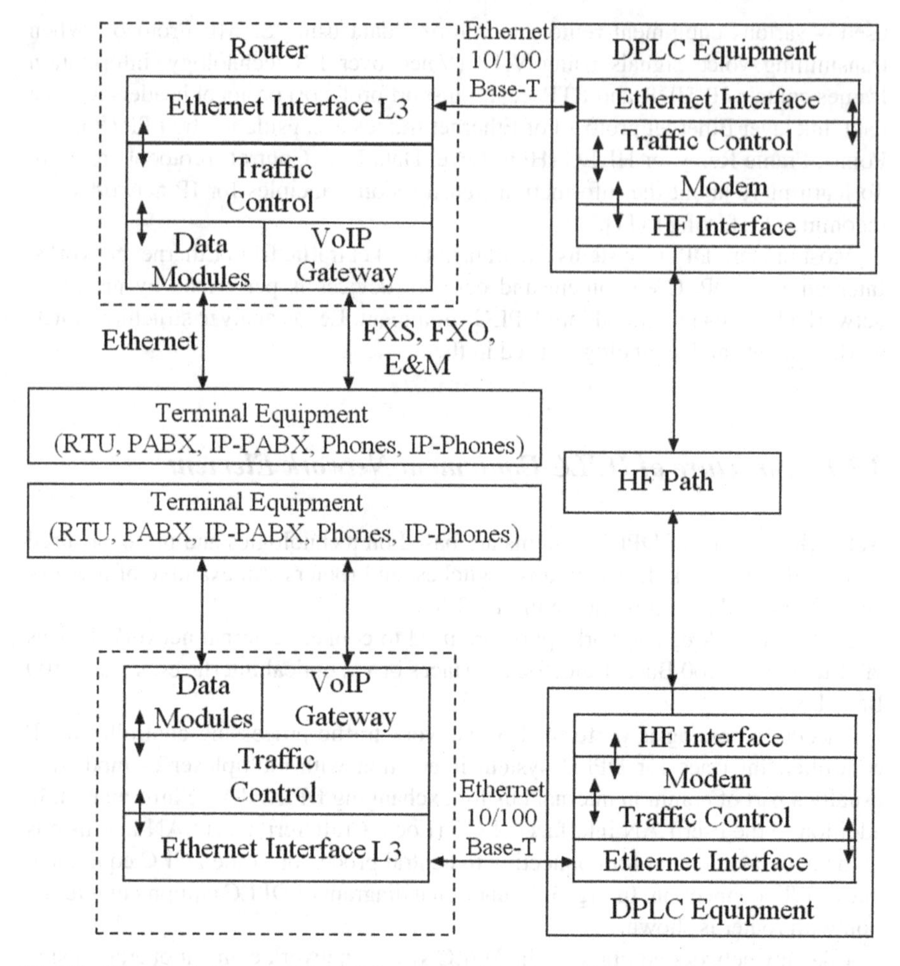

Fig. 3.7 Functional diagram of IP DPLC channel with Ethernet bridge

Table 3.1, seven levels of OSI model are presented, and main network protocols (considering power utilities specifics) which operate at these levels are indicated.

Depending on the network element design, it can either retransmit received traffic in transparent mode or participate in processing of packets when they are transmitted through the DPLC channels. In the first case, two network elements form a transparent Ethernet bridge and DPLC equipment do not participate in packet processing in any way. In the second case, network elements work with data link, network and transport layer information, can perform filtering, prioritization, packet header compression, and packet payload compression.

Network element can also be used for monitoring DPLC equipment using Simple Network Management Protocol (SNMP) and cyber security functions.

Table 3.1 Protocols distribution across seven levels of the OSI model

Layer of OSI model	Protocols
7. Application	General: BGP, DHCP, DNS, HTTP, HTTPs, IPsec, NTP, SNTP, SIP Telecontrol, and power metering: IEC 60870-5-101, IEC 60870-5-104 Voice codecs: G.723.1, G.729 speech codecs, etc. E-mail: POP3, SMTP, File exchange: FTP, SFTP, TFTP Remote access: Telnet
6. Presentation	TLS, SSL
5. Session	H.245, L2TP, PAP, PPTP, RPC, RTCP, SCP, SSH
4. Transport	DCCP, ESP, RTP, RUDP, SCTP, TCP, UDP
3. Network	ARP, EIGRP, ICMP, IGMP, IPv4, IPv6, IPSec, RARP, RIP, OSPF,
2. Data link	ATM, Ethernet, Frame Relay, HDLC, PPP, PPPoE, X.25
1. Physical	Interfaces: Ethernet, RS-232, RS-485, V.35, X.21

3.2.2 *Filtering and Prioritizing Network Traffic*

Packet filtering is used to prevent certain types of traffic, such as broadcast requests, from being transmitted over a low bit rate DPLC channel. DPLC device can filter traffic based on specified criteria, such as the presence in the packet of specific protocol headers.

Packet prioritization is performed with the use of various queue processing disciplines. In hardware buffers of physical LAN interfaces, packet processing is performed using the simplest FIFO (First Input First Output) discipline. In other words, packets do not change their order before they are sent to the Ethernet/IP traffic controller. In the traffic controller, priority queue PQ discipline (Priority Queue) or LLQ discipline (Low Latency Queue) are enabled.

Principle of the priority queue discipline operation is shown in Fig. 3.8. In total, four queues are processed with fixed priorities of processing classes: high, medium, normal, and low. PQ discipline provides guaranteed processing quality for high-priority traffic, but low-priority traffic will be blocked if the channel load is high.

LLQ discipline is a combination of PQ disciplines for high-priority traffic and weighted fair queuing (WFQ) for all other applications, up to 256 queues in total.

Principle of operation of LLQ method is shown in Fig. 3.9.

A distinctive difference of LLQ from PQ is that a certain percentage or specified channel bandwidth is assigned to high-priority traffic.

LLQ discipline is implemented quite simply. Real-time VoIP traffic packets are processed in the priority queue. Data signals, regardless of their priority, are transmitted with the same priority.

PQ discipline is the most suitable for use in DPLC equipment, since it allows setting a strict priority for traffic according to processing class. For example, voice packets are processed in high-priority queue. In the medium priority queue, telecontrol data packets are processed. In the normal priority queue, there are data packets of power metering, and in the low-priority queue, there are data packets of all low-priority optional services.

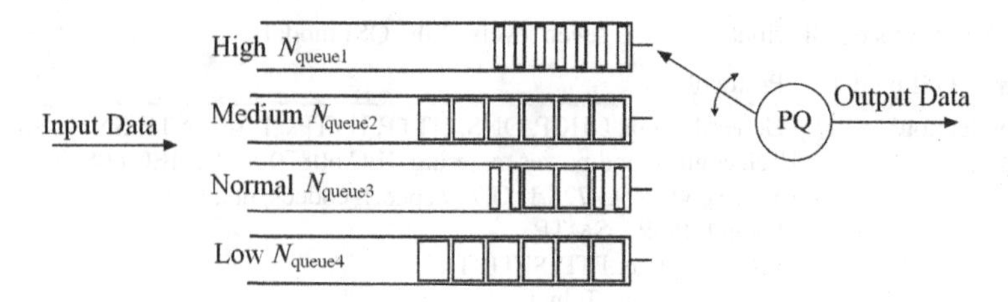

Fig. 3.8 Principle of operation of PQ queue processing discipline

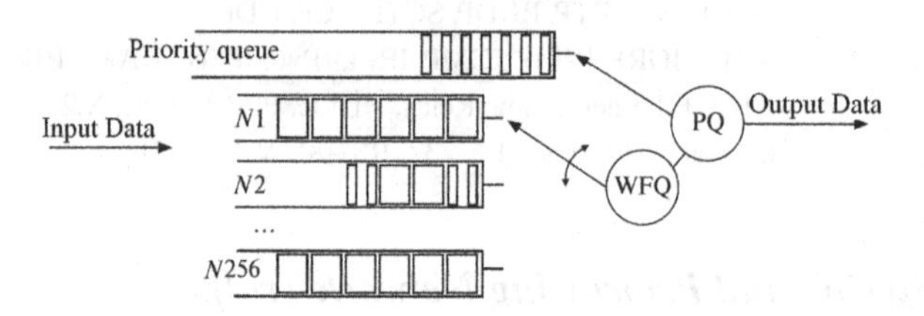

Fig. 3.9 Principle of operation of LLQ queue processing discipline

Complexity of PQ discipline implementing is that the DPLC system network element should differentiate packets of different data signals by some characteristics. For example, the network element should determine type of payload in the packet. If these are IEC 60870-5-104 protocol data, then this data packet originates from telecontrol or power metering systems, and it should be processed in the medium priority queue. If the information sender, for example, remote telemetry unit (RTU), computer, and so on, has a static IP address, the network element of DPLC equipment can decide to distribute packets in queues based on this information. Accordingly, the network element should have an adjustable table of IP addresses of possible data sources.

3.2.3 Header Compression

Information frames in any packet switching network able to switch packets consist of fields of payload and overheads. In IP networks, for some types of traffic, total size of the headers in frame can be several times larger than size of payload. When transmitting data, structure of information frame depends on protocols used – TCP or UDP. For guaranteed delivery according to TCP protocol, information frame has the following structure: 26 bytes, – Ethernet header; 20 bytes, IPv4 header; and 20 bytes, TCP header. For non-guaranteed data transfer, frame structure will be the same, but instead of TCP protocol, UDP protocol is used with a header size of 8 bytes.

For example, information frame with VoIP packet has the following structure: 26 byte, Ethernet header; 20 byte, IPv4 header; 8 bytes, UDP header; and 12 bytes, – RTP header. Total size of the headers will be 66 bytes. Size of payload in VoIP packets is usually 20–30 bytes.

In general, payload size varies depending on settings of application which transmits the information, but it cannot exceed maximum size of Ethernet MTU: maximum transmission unit of 1500 bytes, not considering the Ethernet header itself. In low bit rate channels, to reduce queue waiting time for servicing high-priority packets, MTU size should be configured within a few hundred bytes. Accordingly, total size of the data packet headers will be commensurate with the size of payload. When sending TCP ACK acknowledgment with no payload at all, or with several bytes of information, the DPLC channel bandwidth will only be used for transmitting packet headers, and channel utilization efficiency will be extremely low.

In this section, we do not consider absolute speech or data signal bit rates. We will discuss primary issue of reducing the overhead traffic in information frames. We will consider Ethernet and IP header compression techniques applied to DPLC equipment.

Any header compression techniques work based on the fact that only a small part of the fields in the packet headers change after the session is established. Most of the fields remain static or change small throughout the session. When a session is established, the sender and recipient exchange the session context: information about all the fields in the IP packet header. When using header compression, the sender side performing compression is a compressor, and the receiver side performing header recovery is a decompressor. After exchanging the context, compressor only needs to transmit the difference in the values of the changing fields of two neighboring packets. When getting information about these changes, decompressor can recover the original header. Reliability of the header recovery is checked using a checksum.

If decompressor loses the session context and cannot restore the packet header, it should send a request to compressor to update the context of the session (context state) and wait for the packet with a full-size header to restore the context. However, during round trip time (RTT), decompressor will discard all received packets of the session whose context was lost.

Information about changes in field values can be transmitted by several methods. The simplest way is to send absolute values of changed fields. Moreover, in terms of noise immunity, this method is the most simple and perfect. This is because the fact that when the headers are restored, the packets are independent from each other. Even if packet groups are lost, decompressor can restore headers, since the compressed header contains all information about values of the changed fields.

Another method is to transmit data about the difference between values of header fields of neighboring packets – DELTA encoding. Increase in compression ratio is the DELTA encoding advantage. The disadvantage is that in the header, recovery process packets depends on each other. Packet loss can cause session context damage, since with no data on changing the n-th packet Delta values; the decompressor will not be able to restore the $n + 1$ packet header.

Header compression techniques based on DELTA encoding:

RFC 1144 VJHC (Van Jacobson Header Compression) [5]. This technique makes it possible to compress IPv4/TCP headers from 40 to 4 bytes. Low noise immunity, no mechanism for recovering the session context.

RFC 2507 IPHC (IP Header Compression) [6]. This technique makes it possible to compress IPv4/TCP data packet headers from 40 to 4–7 bytes and compress IPv4/UDP data packet headers from 28 to 2–5 bytes. Improved noise immunity in comparison with VJHC, using periodic refresh of the session context with a specified time interval or number of packets transmitted. Added algorithm for requesting session context updates when the context is damaged. The technique can save the session context using TWICE algorithm when single packets are lost. The algorithm operation is based on the prerequisite that the difference in the header field values for neighboring packets is the same. The decompressor will lose context not when it receives error-damaged packet n, but when it restores, the header of packet $n + 1$. Checking checksum of TCP or UDP, transport protocol will show an error because the difference in the values of the $n + 1$ packet header fields changed by more than specified value (e.g., 1). TWICE algorithm allows increasing the field values by this amount. If repeated calculation of the transport protocol checksum is correct, the session context is considered restored; otherwise the session context needs to be updated by sending a packet with a full header.

RFC 2508 cRTP (Compressed Real Time Protocol) [7]. This is an adaptation of algorithms used in RFC 2507 IPHC technique for transmitting speech packets. IPv4/UDP/RTP compressed header has a minimum size of 2 bytes. When transmitting UDP checksum values, the header size increases to 4 bytes.

RFC 3545 ECRTP (Enhanced Compressed RTP) [8]. Transmitting absolute values of the fields instead of data about the difference thereof. Packets can be transmitted in NO-DELTA mode. In this case, all transmitted fields with changed values will be expressed in absolute values. In NO-DELTA mode, the session context damage is not possible, because even if a packet group is lost, any correctly received packet will have all the data to update the context. In addition, when some field values are transmitted in NO-DELTA mode and the remaining values are transmitted with DELTA encoding, operation in mixed mode is possible. The price of high noise immunity is increase in compressed header size to 12 bytes when the compressor is operating in NO-DELTA mode and to 6–10 bytes when operating in mixed transmission mode.

The most advanced method for transmitting information about changes in header field values is WB-LSB encoding (window-based least significant bit encoding). Compression techniques which use WB-LSB encoding have two main features. First, decompressor can save the session context if number of consecutive lost packets does not exceed the window size. Therefore, such compression techniques can be successfully used in wireless communication channels where radio signal fading is not rare. The second feature is that compressed headers do not transmit difference between changing field values in neighboring headers but information about on change of the lower bits of fields whose values should be transmitted. This approach allows increasing compression ratio while maintaining high noise immunity.

Header compression techniques based on WB-LSB encoding:

RFC 3095 ROHC (Robust Header Compression) [9] and RFC 5225 ROHCv2 [10]. ROHC technique was developed specially for communication channels with high error rates, high transmission delay, and nonsequential packet reception. ROHC technique is multiprotocol and can be used for compression of IPv4/IPv6, UDP, RTP, and ESP protocol headers. The priority ROHC application is compression of headers of audio and video traffic packet. Size of compressed VoIP packet header varies from 1 to 4 bytes depending on whether the header checksum transfer is used.

ROHC v2 is a multiprotocol technique and can process RTP/UDP/IP, RTP/UDP-Lite/IP, UDP/IP, UDP-Lite/IP, and ESP/IP packet headers. In comparison with ROHC, ROHCv2 has several simplifications applied to the compressor and decompressor algorithms. Header recovery algorithm has also been upgraded for high error rates and nonsequential packet reception. ROHC and ROHCv2 techniques are incompatible.

RFC 4996 ROHC-TCP [11]. RFC-TCP header compression technique is based on general algorithms used in RFC RFC 3095. It is used for compression of TCP headers. Size of compressed IPv4/TCP header is 4–7 bytes.

Various disturbances in DPLC channels can cause packet loss. Let us analyze how packet loss can affect decompressor operation:

1. Packets will be lost, and this will not cause the session context damage.
2. Packets will be lost, and decompressor will not be able to save the session context.

Since there was no context loss in the first case, the disturbing factor consequences will be:

1. For VoIP session: short-term loss of voice signal for listening user followed by its recovery after period equal to total duration of the voice signal in the lost packets.
2. For guaranteed data transmitted using TCP protocol: application level initializes retransmission of the lost packets.

If the session context is received, the packet loss consequences can be considered insignificant in both cases.

The consequences will be much more serious if the session context is lost. A very important characteristic of header compression techniques is robustness q_{RB}. It is defined by number of error packets which can be successively received before the session context damage occurs. Initially, header compression techniques which use DELTA encoding have worse noise immunity in comparison with techniques based on WB-LSB encoding. Let us analyze noise immunity characteristics of the compression techniques under consideration:

1. IPHC and cRTP techniques have low noise immunity. TWICE algorithm processes well loss of single packets, but if two or more packets are lost, the session context will be lost.
2. ROHC technique ensures safety of context sessions at number of lost packets up to 62.

Using NO-DELTA mode in ECRTP technique when transmitting values of all changing fields allows not to worry about losing the session context at all, since any correctly received packet will restore it as all changed field values are known. If NO-DELTA IP ID transmission mode is used, then knowing load type in the speech packet and codec characteristics, decompressor can restore RTP timestamp field and continue operation if UDP checksum is correct.

Analyses on the use of IP packet header compression in noisy channels can be found in publications [12–14] are of great interest.

In [15], simulation of VoIP packet transmission with compressed headers is performed using cRTP, ECRTP, and ROHC techniques. As a result of the study, graphs of decompression failure rate versus percentage of lost packets shown in Fig. 3.10 were obtained.

The results confirm low noise immunity of cRTP technique even if TWICE algorithm is applied. Regardless of operating modes, ROHC and ECRTP techniques show high noise immunity even at 20% packet loss. Article [16] presents data on percentage of packet losses at different values of BER obtained as a result of the experiment. During this experiment, speech intelligibility was evaluated when transmitting VoIP packets over DPLC channel, and it was found that even at BER = 10^{-4}, and in the absence of compression of packet headers, quality of speech signal perception was legible, whereas it was almost impossible to understand speech for cRTP, but ECRTP and ROHC provided satisfactory quality of speech perception even in conditions of high error rates 10^{-4}, 10^{-5}.

ROHC/ROHCv2 with EthHC	ROHC-TCP without EthHC	ROHC-TCP with EthHC	ECRTP via PPP	IPHC via PPP
Ethernet	Ethernet	Ethernet	PPP	PPP

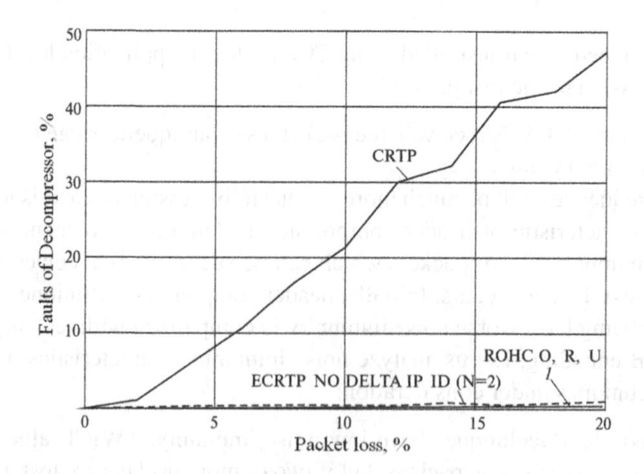

Fig. 3.10 Graphs of decompressor failure rate versus percentage of lost packets. Curves for the techniques: ECTP NO DELTA ($N = 2$), ECRTP, ROHC in O, R, U modes, are almost the same and correspond to curve 1

ROHC/ROHCv2 with EthHC	ROHC-TCP without EthHC	ROHC-TCP with EthHC	ECRTP via PPP	IPHC via PPP
26	26	26	8	8
14	–	14	–	–
IPv4/UDP/RTP	IPv4/TCP	IPv4/TCP	IPv4/UDP/ RTP	IPv4/TCP
40	40	40	40	40
4	7	7	12	7
66	66	66	48	48
18	33	21	20	15
3.67	2.00	3.14	2.40	3.20

Thus, in DPLC equipment, it is recommended to use robust header compression techniques ROHC, ROHCv2, ECRTP for VoIP packets, and ROHC-TCP for TCP packets.

All the above techniques allow reducing size of transport and network layer protocol headers. To reduce size of Ethernet data link layer protocol header, two approaches can be used:

1. To apply Ethernet header compression: Effnet EthHC (Ethernet Header Compression). This technique is developed as an addition to ROHC/ROHC-TCP [17]. Compressed header size without physical layer preamble is 6–8 bytes. But there is one issue, which should be taking into account. Initially Ethernet frame has 8 bytes preamble on physical layer. When processing Ethernet frames, DPLC manufacturers apply various approaches. It could be cutting of preamble and adding some internal service information and checksum to transmit data via DPLC channel. Or it could be cutting of preamble and use of Point-to-Point over Ethernet (PPPoE) tunneling with additional 8 bytes header of Point-to-Point protocol. Therefore, to evaluate compression ratio, we assume some average L2 compressed header size even with application of EthHC of 14 bytes;
2. To use Ethernet frame encapsulation with PPP protocol. In this case, the DPLC equipment network element replaces Ethernet header with PPP (Point to Point) protocol header. PPP header has size of 8 bytes. The use of PPP encapsulation requires a router as the DPLC equipment network element. It should be noted that this approach is used when creating serial WAN connections between two external routers. Routers are connected to DPLC equipment via X.21 synchronous serial interface instead of network element and Ethernet connection. DPLC equipment does not involved in processing of packets in this case, just converting electrical signal from X.21 interface to HF signsl and visa versa. Functional diagram of such a DPLC channel is shown in Fig. 3.11.

In Table 3.2, data on compression ratio of frame headers for various compression techniques are given. As can be seen, header can be reduced by more than three times when using EthHC compression and ROHC/ROHC-TCP techniques.

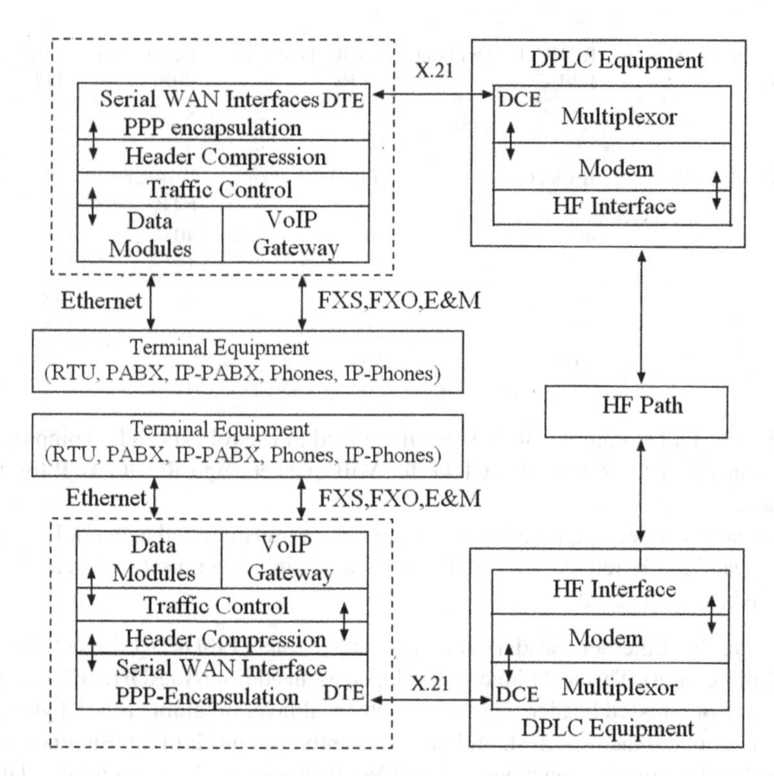

Fig. 3.11 Functional diagram of IP DPLC channel with synchronous serial WAN connection

Table 3.2 Header compression techniques

Header parameter	ROHC/ROHCv2 without EthHC
L2 layer protocol	Ethernet
Size of L2 header before compression, bytes (including preamble)	26
Size of L2 header after compression, bytes	–
L3/L4 layer protocols	IPv4/UDP/RTP
Size of L3/4L4 headers before compression, bytes	40
Size of L3/L4 headers after compression, bytes	4
Total size of headers before compression, bytes	66
Total size of headers after compression, bytes	30
Compression ratio	2.20

3.2.4 Payload Compression of Data Packets

Compression of IP packet headers is aimed at reducing overhead traffic. There are several technologies focused on compression of data frames in general. They are based on Lempel–Ziv–Stac compression algorithm (LZS algorithm) [18]. The use of this algorithm for data stream compression is applied in three main technologies.

Stac compression is frame compression technique developed by Space Communications based on RFC 1974 document PPP Stac LZS Compression Protocol [19].

The compression algorithm uses an encoding dictionary which replaces source data stream with code words. This technology can work both with frames of a single session or with several sessions at once. In STAC Compression algorithm, data compression is performed inside a 2048-byte sliding window formed from incoming data stream. In detail, principle of adaptive encoding with sliding window method is discussed in ANSI X3.241-1994 standard. The compressor searches for a match of input code sequence inside the sliding window and, when it is found, performs encoding according to the code table.

Predictor Compression. Data compression algorithm based on predicting incoming code sequence. This algorithm is described in RFC 1978 PPP [20].

Unlike Stac algorithm, predictor algorithm is based on the original encoding table. The compressor predicts sequences of symbols based on data of previously transmitted frames. Predictor compression algorithm requires from network equipment high computing resources, but it provides very low data processing delay.

Compression control protocol for PPP connections – PPP Compression Control Protocol RFC 1962 [21] is used to control operation of the compressor and decompressor which use Stac and Predictor algorithms.

IP Payload Compression Protocol (IPComp). The technique uses LZS algorithm to compress payload in IP packets. Principle of this technique operation is described in RFC 2393 [22] document. Unlike Stac and Predictor, IP Payload only compresses IP packet load without affecting IP header and layer two frame headers. In this case, IP packets which are involved in transmitting encoded information, 4 bytes of service information are added to IP header. Compressor and decompressor operation is monitored using proprietary IPComp protocol procedures.

Based on LZS algorithm, V. 42 bis compression protocol was developed, which is widely used in low bit rate modems operating on copper communication lines.

Efficiency of LZS-based compression techniques strongly depends on two factors: type of information being compressed and data frame size. Text information is compressed best of all. At that, compression ratio can reach 8:1. Image files and speech which have a compression ratio of less than 2:1 are the worst compressed. The more the IP packet size in the information frame, the more the compression ratio. RFC 2395 document states that the use of IP payload compression techniques is only appropriate if the information size in the IP packet is at least 90 bytes.

Use of LZS compression for DPLC channels is promising when transmitting data which are text information, for example, from substation telecontrol and and power metering systems

3.2.5 *SNMP Monitoring of DPLC Equipment*

Network management system (NMS) allows monitoring various communication equipment using a simple network management protocol, SNMP (Simple Network Management Protocol). Implementation of SNMP monitoring function in DPLC equipment makes it possible to include it in the general enterprise network management system along with any other devices which can transmit information about their status via SNMP Protocol. Principle of SNMP monitoring is shown in Fig. 3.12.

The network management system server is SNMP Manager, and the DPLC equipment is SNMP agent. Basic function of SNMP Manager is to poll status of SNMP agents. Interaction between NMS server and the device being polled is performed using MIB tables (management information base).

The table contains information about status of the device being polled and about emergency and warning messages. After MIB file compiling, management system can request for status of the DPLC equipment network element at specified IP address. In turn, DPLC equipment can also transmit traps about changes in its state, such as accident, without waiting for polling by server, via spontaneous messages. To do this, IP address of the NMS server is entered in the DPLC network element configuration.

Data transfer from SNMP agent can be performed using different versions of SNMP protocol. Access to SNMP agent is protected by a password. In SNMPv2 protocol, transmission of password information to the agent is implemented in unsecured mode, not using encryption. This is a serious problem from the point of view of cyber security. The password can be successively attacked, which means full access to SNMP agent. To protect connection with SNMP agent and transmitted data, the third version of SNMP protocol – SNMPv3 is used. When using SNMPv3 protocol, cyber security is provided by the following methods:

1. Security model based on user USM (user-based security model). In this case, each user has a unique name, authorization, and privacy key. When transmitting authorization data, MD5 and SHA-1 encryption protocols are used. SNMP agent

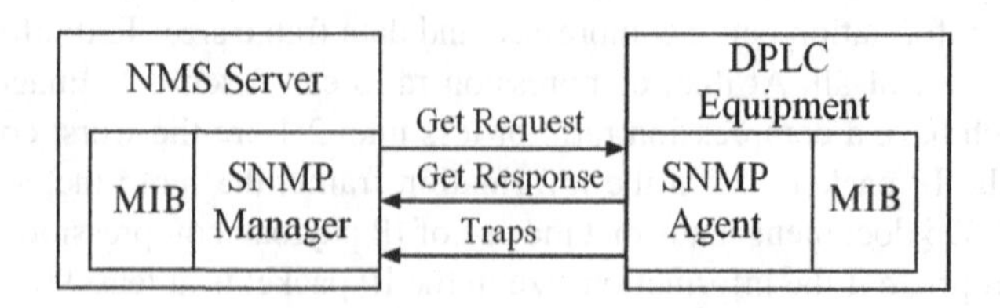

Fig. 3.12 Diagram of interaction between SNMP Manager and SNMP agent

checks authorization key when it receives a request and resets the connection if the authorization is incorrect. The authorization key itself is encrypted and decrypted using DES (data encryption standard) symmetric encryption algorithm.
2. Access control model based on VACM view (view-based access control model). Is used to control user access to information MIB table objects.

References

1. Olifer, V., Olifer, N.: Computer networks: principles, technologies, and protocols for network design. Wiley (2006)
2. Coding of speech at 8 kbit/s using conjugate-structure algebraic-code-excited linear-prediction. Recommendation G.729, Telecommunication Standardization Sector of ITU, Geneva, Switzerland (1996)
3. Dual rate speech coder for multimedia communications transmitting at 5.3 and 6.3 kbit/s. Recommendation G.723.1, Telecommunication Standardization Sector of ITU, Geneva, Switzerland (1996)
4. Romanov, S.E.: Design of DPLC channels and networks (in Russian) (2005). http://romvchv-lcomm.blogspot.com/2007/08/blog-post_8270.html. Accessed 01 June 2020
5. Jacobson, V.: Compressing TCP/IP headers for low-speed serial links. RFC 1144, IETF (1990)
6. Degermark, M., Nordgren, B., Pink, S.: IP header compression. RFC 2507, IETF (1999)
7. Casner, S., Jacobson, V.: Compressing IP/UDP/RTP headers for low-speed serial links. RFC 2508, IETF (1999)
8. Koren, T., Casner, S., Geevarghese, J., Thompson, B., Ruddy P.: Enhanced compressed RTP (CRTP) for links with high delay, packet loss and reordering. RFC 3545, IETF (2003)
9. Bormann, C.: Robust header compression (ROHC) over PPP. RFC 3241 (2002)
10. Pelletier, G., Sandlund, K.: Robust header compression version 2 (ROHCv2): profiles for RTP, UDP, IP, ESP and UDP lite. RFC 5225, IETF (2008)
11. Pelletier, G., Sandlund, K. Jonsson, L.-E., West, M.: Robust header compression (ROHC): a profile for TCP/IP (ROHC-TCP). RFC 4996, IETF (2007)
12. Degeermark, M., Hannu, H., Jonsson, L., Svanbro, K.: Evaluation of CRTP performance over cellular radio links. IEEE Pers. Commun. **7**(4), 20–25 (2000)
13. Fitzek, F., Rein, S., Seeling, P., Reisslein, M.: Robust header compression (ROHC) performance for multimedia transmission over 3G/4G wireless networks. Wirel. Pers. Commun. **32**, 23–41 (2005)
14. Joseph, A., Ishac, G.: Survey of header compression techniques. Glenn research centre, Cleveland, Ohio, NASA/TM 2001-211154 (2001)
15. Knutsson, C.: Evaluation and implementation of header compression algorithm ECRTP. Master Thesis. Lulea University of Technology (2004)
16. Merkulov, A.G.: Experimental research of VoIP speech transmission along DPLC channels (in Russian), SIBSUTIS, Novosibirsk №3 45-56 (2014)
17. An introduction to IP header compression. White paper. Effnet AB. http://www.effnet.com/pdf/Whitepaper_Header_Compression.pdf (2004). Accessed 01 June 2020
18. ANSI X.3241-1994 Data compression method – adaptive coding with sliding window for information interchange. American National standard for information systems. American National Standards Institute, Inc. (1994)
19. Friend, R., Simpson, W.: PPP Stac LZS compression protocol. RFC 1974, IETF (1996)
20. Rand D.: PPP predictor compression protocol. RFC 1978 IETF, (1996)
21. Rand D.: The PPP compression control protocol (CCP). RFC 1962, IETF (1996)
22. Shacham, A., Monsour, R., Pereira, R., Thomas, M.: IP payload compression protocol (IPComp). RFC 2393, IETF (1998)

Chapter 4
Features of Frequency Characteristics of the HF Paths

The main parameters of the HF path are its attenuation, return loss, and level of noise. These parameters have the greatest impact on operating conditions of DPLC communication channels (at a given output power, attenuation of the path determines reception level of transmitted DPLC signals, and return loss defines deviation of the HF path input impedance from its nominal value and, accordingly, operation conditions of the PLC equipment transmitter). All results of HF path and line path frequency response calculations given in this chapter and Chap. 5 are obtained by modeling using WinTrakt specialized software. Brief information about this application is provided in Annex 1 to this book. WinTrakt and WinNoise software are widely used by engineering companies in various countries, as well as by manufacturers of PLC equipment.

DPLC device is terminal equipment (TE) installed at the ends of the communication line connecting TE sets to each other. In the terminal equipment, information signals received by the user interfaces at one end of the channel are converted into a form intended for transmission over the communication line and reverse conversion of signals received from the communication line to send them to the user interfaces at the other end of the channel. The communication line is the medium through which converted signals are transmitted between the terminal equipment.

PLC channels relate to wired communications which use metal conductors of electric power transmission lines as the medium for transmitting signals. Block diagram of PLC channel indicating component parts is shown in Fig. 4.1.

PLC communication channels use phase conductors and insulated groundwires of overhead electric power transmission lines (OPL), as well as phase conductors and sheaths of underground power transmission cable lines (CPL) as a medium for signal propagation.

Let us remind that the PLC channel consists of HF path and terminal equipment. The HF path consists of line path (overhead or cable power line, including tap lines, transpositions, HF bridges), line traps, coupling capacitors or capacitive voltage transformers (CVT), coupling devices (line matching units), and HF cables.

A. G. Merkulov et al., *High Voltage Digital Power Line Carrier Channels*,
https://doi.org/10.1007/978-3-030-58365-1_4

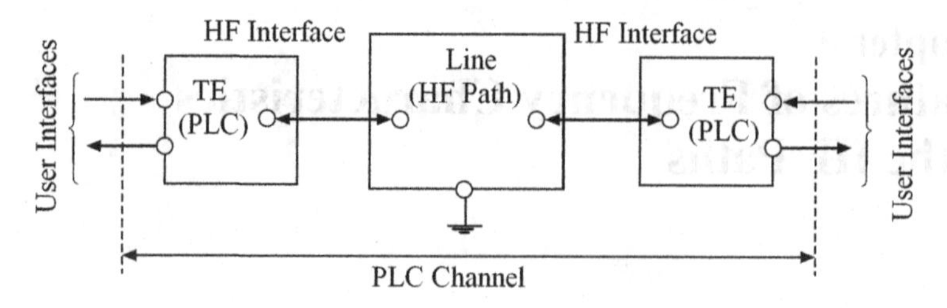

Fig. 4.1 Block diagram of PLC channel

In conventional wired communications, the communication line design is especially developed to obtain optimal parameters for transmitting signals over them in the frequency range under consideration (well matching of impedances, high cross talk attenuation between channels on different lines, relatively low level of noise, etc.).

Electric grids formed by high-voltage substations (SS) and power lines are originally intended to transmit electric power. To transmit communication signals, they must be adapted to a certain extent using special devices. Therefore, the HF path parameters (this term is used in PLC to call communication line) are much worse than similar parameters of other types of wired communications.

Because of this, the creation of PLC channels using power lines and parameters of these channels significantly differ from channels of conventional wired communication lines.

These differences consist in the fact that in the PLC channels over a power line:

- Special equipment is used which is intended for:
  - Connection to high-voltage power lines. These devices are coupling capacitors or CVT and coupling devices (line matching units).
  - To separate HF path through which signals are transmitted from shunting action of input impedance of SS between which the power lines are connected and input impedance of the tap lines. These devices are line traps (LT) with different schemes of tuning elements.
- There are specific wideband disturbances of quite high level:
  - Permanent (due to corona on conductors and electric discharges in power line insulation).
  - Short-term impulses (overvoltage due to lightning, switching, and short circuits in power lines).
- Signals transmitted in the channel over a power line are distributed over the electric grid far beyond the limits of this power line. Operability of each channel operating in common electric grids should be determined considering the presence of these interfering signals.
- Conditions for signal propagation over power lines differ significantly from conditions for other types of wired communication lines.

This is caused by:

- Design features of OPL (distances between conductors and between conductors and earth are close that leads to significant penetration of the overhead line electromagnetic fields in the earth and losses in the earth).
- Impossibility to ensure matching of PLC output and OPL impedances at the ends of homogeneous sections of overhead lines (reflection coefficient value can reach 0.7).
- Presence of tap lines.
- Multimode HF signal propagation over power lines. As different modal components involved in the signal propagation have different propagation velocity, attenuation poles appear on the frequency response of the HF path (increase in the HF path attenuation near certain frequencies).
- Effect of coupling (to which the coupling is made) and "non-coupling" phase load impedance on HF signal propagation conditions. This impedance depends on the power line switching state, and when the power line is energized, it depends on input impedance of substation to which the power line is connected.
- Air temperature which affects the overhead lines' phase and cable slack and, accordingly, value of earth losses.

All this can cause quite complex frequency dependence of HF path parameters, and the parameters themselves can be time-unstable. These features of power lines as communication lines should be taken into account when drawing up technical specifications for the PLC equipment development, designing PLC channels, and at these channels' operation.

4.1 Features of Attenuation Frequency Dependences of the HF Paths

When analyzing the features of attenuation frequency dependence, we will use a modal theory of wave propagation along the multiconductor homogeneous line above the earth.

The basics of the modal theory itself and method for calculating parameters of power line modes are described in [1–10] and will not be considered here. If desired, readers can know more about them in publications. Note only that numerical values of modal parameters for power lines of any design can be calculated using WinTrakt program [11].

However, before analyzing features of attenuation frequency dependences for which we will use the modal theory, we will give a brief description of power line modal parameters with examples of these parameters' numerical values for specific structure power lines and will consider a method used for this theory to analyze frequency response of the HF path attenuation.

Let us remind that the HF path is characterized by HF path attenuation, which is the sum of line path attenuation, line trap losses, and coupling losses.

4.1.1 Brief Information About Modal Parameters

In the framework of the modal theory, any homogeneous section of a power line is characterized by modal parameters defined in a special way, described in matrix form. These parameters are:

λ – square complex n-th order matrix for transformation of a system of mode voltage vectors into a system of phase voltage vectors.
δ – square complex n-th order matrix for transformation of a system of mode current vectors into a system of phase current vectors.
$\gamma = \alpha + j\beta$ *is* a diagonal complex n-th order matrix of modal propagation constant (real part α is attenuation constant, and imaginary part β is phase constant).
$Z_{m.w.}$ is a diagonal complex n-th order matrix of modal wave impedances.

The λ and δ matrices are related by the following formula:

$$\delta = \left(\lambda^{-1}\right)'$$

The phase constant β of each mode is related to the velocity of this wave mode propagation over the power line υ. This relationship is determined by a formula which can be written as (for k-th mode):

$$\beta_k = \frac{2\pi f}{\upsilon_k} \tag{4.1}$$

where $\kappa = 1,2\ldots n$.

In an ideal lossless line, the attenuation constant is zero, and the propagation velocity is equal to light speed.

The difference between a propagation constant of each mode and its value for lossless line is determined by the losses in the conductors and the earth for this mode.

The modes are usually divided into two main groups. The first includes so-called earth (or zero) mode with maximum attenuation constant and maximum difference between propagation velocity and light speed (for air $3 \bullet 10^8$ m/s).

Other $(n - 1)$ modes are called interphase modes. Numbers from 1 to $(n - 1)$ are usually assigned to them, considering their ranking by the degree of increasing losses occurred in these modes. The less the propagation loss, the smaller number of mode. It should be noted that the difference in insertion losses of each mode is mainly determined by the difference in losses caused by the earth.

Values of elements of λ, δ, and $Z_{m.w.}$ matrix depend on frequency relatively little. Values of elements of $\gamma = \alpha + j\beta$ matrix, as can be understood from the above text, depend on frequency significantly.

To get an idea of the modal parameter values, we will give these values for a three-phase 220 kV overhead line with a horizontal phase arrangement that were obtained by calculating at a specific soil resistivity of 100 ohm•m. A drawing of this overhead line tower is shown in Fig. 4.2.

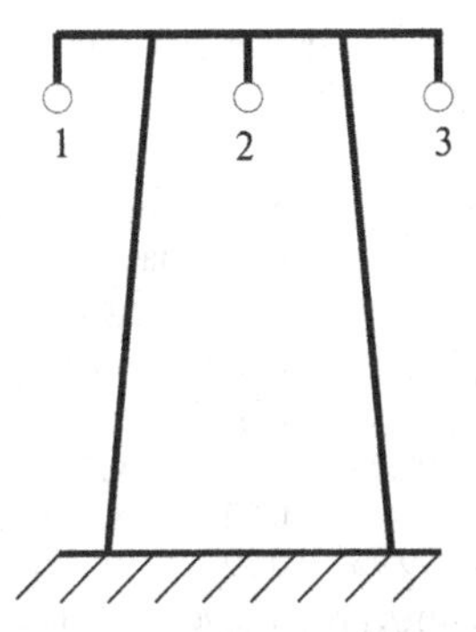

Fig. 4.2 Drawing of overhead line tower with horizontal phase arrangement

Further, when analyzing the frequency response shape for attenuation of the HF path, we will use this OPL; since it is for OPL with a horizontal phase arrangement, all features of frequency characteristics of different coupling schemes and different environmental conditions are most clearly obvious. At that, it is meant that, for overhead lines with other types of towers, all arguments and conclusions about features of frequency characteristics of the HF path attenuation remain valid.

Values of matrix elements λ, δ, and $Z_{m.w.}$ for the overhead line Fig. 4.2 (frequency 100 kHz) are given as:

$$\lambda = \begin{pmatrix} 1;0 & 1;0 & 1;0 \\ 2,09;3,124 & 0;0 & 1,03;0,011 \\ 1;0 & 1;3,14 & 1:0 \end{pmatrix}$$

$$\delta = \begin{pmatrix} 1;0 & 1;0 & 1;0 \\ 1,94;3,131 & 0;0 & 0,96;0,018 \\ 1;0 & 1;3,14 & 1;0 \end{pmatrix}$$

Note: The 1st and 2nd columns of the matrices λ and δ belong to the interphase modes (1 and 2, respectively) and the last column to zero (earth) mode.

The diagonal complex matrix of n-th order for mode wave impedances will have the form:

$$Z_{m.w.} = \left[(342.6;0.003);(411.8;\ 0.004);(619.9;\ 0.046)\right]$$

Note: elements of matrices λ, δ, and $Z_{m.w.}$ are written in the following form: (module, phase angle, radian) the diagonal matrix $Z_{m.w.}$ for convenience, is represented in the form of a line, rather than a diagonal matrix.

To represent elements of the propagation constant matrix $\gamma = \alpha + j\beta$ for 220 kV overhead line under consideration and their dependence on frequency, the values of mode attenuation constants α, phase constants β, and propagation velocity υ related with phase constants β by the formula (4.1) are given in Table 4.1.

Table 4.1 Attenuation constant α, phase constant β, and propagation velocity υ for the modes of the 220 kV overhead line under consideration

Frequency, kHz	Mode name					
	Mode 1 center phase-to-outer phases		Mode 2 outer phase-to-outer phase		Mode 0 Earth mode	
	α	β/υ	α	β/υ	α	β/υ
50	0.023	1.0501/299.17	0.035	1.0578/296.99	0.56	1.162/270.36
100	0.033	2.0987/299.38	0.065	2.1123/297.46	0.95	2.2704/276.74
200	0.047	4.1951/299.54	0.12	4.2183/297.9	1.57	4.452/282.26
400	0.069	8.387/299.7	0.22	8.4242/298.3	2.52	8.7632/286.8
800	0.1	16.769/299.75	0.39	16.826/298.74	3.92	17.312/290.35

Notes: units of matrix elements: α (dB/km), β (rad/km), υ (thousand km/s)

The difference between the values of attenuation constants and propagation velocities of different modes, due to different earth losses for these modes (for mode 1, the earth losses are minimal; they are practically absent), is clearly shown from the data given in Table 4.1, and we can say that the attenuation constant and deviation of the propagation velocity from light speed are caused only by losses in the phase conductors.

All modes' attenuation constants increase monotonously with increasing frequency.

The voltages and currents conversion from modal to phase ones and reverse conversion can be written in matrix form as follows:

$$U_{\text{ph}} = \lambda U_{\text{m}}; I_{\text{ph}} = \delta I_{\text{m}}; \tag{4.2}$$

$$U_{\text{m}} = \lambda^{-1} U_{\text{ph}}; I_{\text{m}} = \delta^{-1} I_{\text{ph}}; \tag{4.3}$$

where U_{m} and I_{m} are n-order column complex matrices of mode voltages and currents and U_{ph} and I_{ph} are n-order column complex matrices of phase voltages and currents.

For an incident wave, the relationship between mode voltages (currents) at the beginning ($U_{\text{m.b}}$) and end ($U_{\text{m.e}}$) of the line section under consideration with length L (km) is written for each r-th mode ($r = 1, 2,... n$) in the following form:

$$\begin{aligned} U_{\text{m}r\text{.e}} &= \exp\left(-\left(\alpha + j\beta_r\right)L\right) U_{\text{m}r\text{.b}} \\ I_{\text{m}r\text{.e}} &= \exp\left(-\left(\alpha + j\beta_r\right)L\right) I_{\text{m}r\text{.b}} \end{aligned} \tag{4.4}$$

Note: it should be remembered that the attenuation constant α (real part of the propagation constant γ) in the formula (4.4) should be expressed in Neper/km and the phase constant β should be expressed in rad/km.

When defining mode voltages (currents) at the end section of OPL, we, as a rule, are interested not in absolute values of the phase angles βL but only in the phase shift between the voltages (currents) of different modes relative the mode phase, considered as the main (typically mode 1). Therefore, the formula (4.4) is often written as (for r-th mode):

$$\begin{aligned} U_{mr.e} &= \left(\exp(-\alpha L)\left(\cos(-\Delta\beta_r L)\right) + \sin(-\Delta\beta_r L)\right) U_{mr.b} \\ I_{mr.e} &= \left(\exp(-\alpha L)\left(\cos(-\Delta\beta_r L)\right) + \sin(-\Delta\beta_r L)\right) I_{mr.b} \end{aligned} \tag{4.5}$$

where for r-th mode:

$$\Delta\beta_r L = \frac{2\pi fL}{\upsilon_1 \upsilon_r}(\upsilon_1 - \upsilon_r) = \frac{2\pi fL}{\upsilon_1 \upsilon_r}\Delta\upsilon_{1r} \tag{4.6}$$

Here, frequency is expressed in kHz, velocity – in thousand km/s and length L – in km.

The essence of using the modal theory in analyzing features of attenuation frequency dependences for HF path can be understood from Fig. 4.3.

In the original line model (Fig. 4.3a), the propagation of incident voltage wave along the line occurs in phases which have mutual attenuation per unit length. The presence of this mutual attenuation significantly complicates wave propagation along the line introducing dependence of a phase current and voltage on currents and voltages of all other phases.

In the converted model (Fig. 4.3b), at the beginning of the line, the transition from phase voltages to mode voltages is performed using transformation (4.3). The incident wave modal voltage propagation along the line from its beginning to its end occurs for each mode independently of other modes along certain four-pole network whose parameters γ and $Z_{m.w.}$ correspond to the mode parameters. Then, using the transformation (4.2), the transition to incident wave voltages at the phases at the end of the line is performed.

Thus, the process of describing incident wave propagation over a homogeneous section using a converted line model consists of a sequence of the following actions (with the example of voltages of a three-phase overhead line):

Step 1. Using the first of the Eq. (4.2), a given system of n phase voltage vectors at the beginning of the section (matrix $U_{ph.b.}$), determined by conditions at the end, for the beginning of the considered power line section (signal sources connected to the phase conductors and each phase load), is converted into a system of n modal voltage vectors (matrix $U_{m.b.}$).

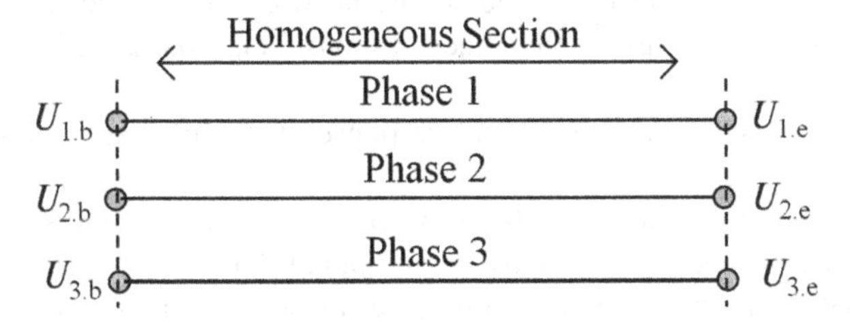

Fig. 4.3 Principle of a real overhead line replaced with a model with modal components: (**a**) original model and (**b**) converted model

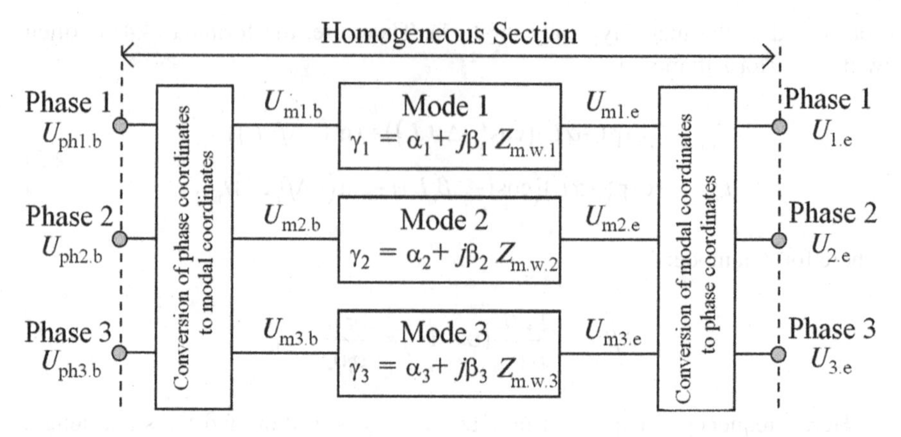

Fig. 4.3 (continued)

Step 2 Using the first of the Eq. (4.2), determine each mode voltages at the end of the considered section.

Step 3 Using the first of the Eq. (4.3), a system of n mode voltage vectors (or current) at the end of a homogeneous section of the power line is transformed into a system of n phase voltage (or currents) vectors (matrix $U_{ph.e.}$).

4.1.2 Frequency Response of the Line Path Without Considering Multiple Reflected Waves

In this section, we will see how the line path frequency response is affected by the phase selected to couple to the line. The analysis will be carried out under the assumption that reflected waves which "obscure" the nature of the line path frequency response are absent. The effect of reflected waves will be considered separately.

First, we will present the results of calculating frequency response of the line path according to the scheme shown in Fig. 4.4. The calculation was performed using the WinTrakt software.

The power line included in the HF path is a three-phase 220 kV overhead line with a horizontal phase arrangement for which modal parameters were given above. The OPL length is 80 km.

This scheme with the specified end conditions was selected because this scheme is most convenient for further frequency response analysis using modal analysis methods.

In Fig. 4.5, the calculated frequency response of the line path of this OPL, for different coupling schemes, is presented.

To understand the causes of this type of frequency characteristics given in Fig. 4.5, we use the above method to describe signal propagation along the line with application of the modal method. In this case, we will use numerical values of the modal parameters for the considered overhead line given above.

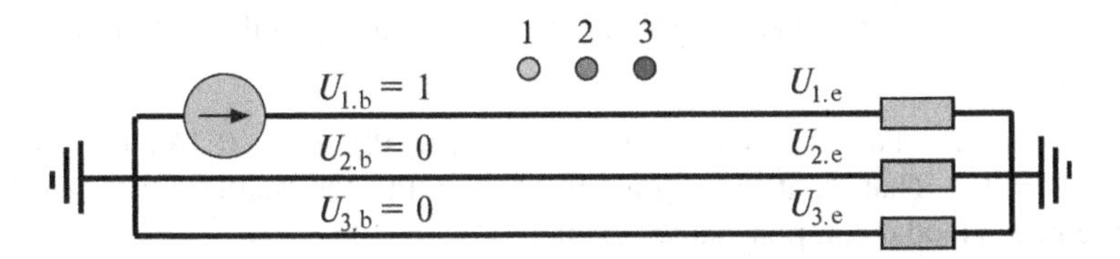

Fig. 4.4 Analyzing the conditions of signal propagation along a homogeneous OPL section (example of connecting the signal source to phase 1)

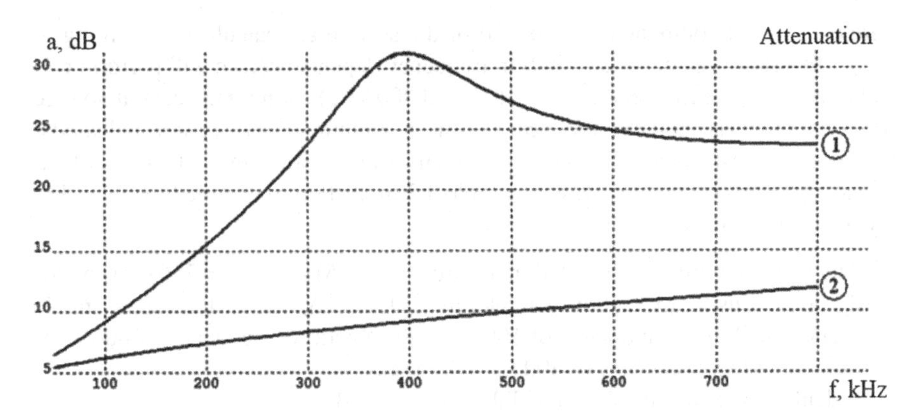

Fig. 4.5 Frequency response of the line path for various phase-to-earth coupling schemes. *Graph 1* outer phase-to-earth, *graph 2* center phase-to-earth

We will make assumptions which do not affect the essence of the analyzed processes but significantly simplify their consideration. These assumptions are as follows:

- Matrices λ and δ and their inverse matrices are not represented by their values given above, obtained by calculation, but represented in a simplified form (purely real and with rounded values of elements):

$$\lambda \approx \delta \approx \begin{pmatrix} 1 & 1 & 1 \\ -2 & 0 & 1 \\ 1 & -1 & 1 \end{pmatrix}; \lambda^{-1} \approx \delta^{-1} \approx \begin{pmatrix} 1/6 & -1/3 & 1/6 \\ 1/2 & 0 & -1/2 \\ 1/3 & 1/3 & 1/3 \end{pmatrix}$$

- Each phase at the receiving end of the overhead line is loaded with an impedance equal to the input impedance of the line path.

Let us consider different options for connecting the transmitter and receiver to the overhead line when making calculations according to the steps described above.

Under specified end conditions, the column matrix of relative phase voltages at the beginning of the section can be written as

- For phase 1-to-earth coupling $U_{ph.b} = |1; 0; 0|$.
- For phase 2-to-earth coupling $U_{ph.b} = |0; 1; 0|$.

Note: For convenience, the column matrix $U_{ph.b}$ is written as a row, not a column.

Stage 1 Modal components are calculated at the beginning of the section according to (4.3):

- For coupling phase 1-to-earth (at $U_{ph.b} = |1;\ 0;\ 0|$): $U_{m1.b} = 1/6$; $U_{m2.b} = 1/2$; $U_{m0.b.1} = 1/3$. That is, at the beginning of the OPL, the "earth" mode and both interphase modal components are excited.
- For connection phase 2-to-earth (at $U_{ph.b} = |0;\ 1;\ 0|$): $U_{m1.b} = -1/3$; $U_{m2.b} = 0$; $U_{M0.b.1} = 1/3$. That is, at the beginning of the OPL, in addition to the "earth" mode, only one interphase mode (the first one) is excited.

Stage 2 Modal components at the end of the section are calculated according to (4.5): To get an idea of the attenuation frequency dependence, we will perform calculations for three frequencies (100, 400, and 800 kHz) taking into account voltage values of various modal components at the beginning of the section and modal attenuation constants and propagation velocity values shown in Table 4.3. At that, the attenuation constant expressed in Table 4.3 in dB/km is converted to Neper/km by multiplication by 0.115.

The value required for calculation according to (4.5) $\Delta\beta_r L$ ($r = 2; 0$) was previously defined using (4.6). For example, the value of $\Delta\beta_r L$ for the second mode ($r = 2$) is equal to for frequency of 100 kHz, 1,088 rad; for frequency of 400 kHz, 2,976 rad; and for frequency of 800 kHz, 4,56 rad.

Calculation results are shown in Tables 4.2 and 4.3.

Stage 3 The voltage of each phase at the end of the section is calculated according to (4.5). Calculation results for step 3 are given in Table 4.3.

Table 4.2 Modal components at the end of the line with different transmitter coupling schemes at the beginning of the line

Frequency, kHz	Modal components at the end of line for coupling scheme	
	Phase 1-to-earth	Phase 2-to-earth
100	$U_{m1.e} = \exp(-0{,}115 \cdot 0.033 \cdot 80)/6 = 0{,}123$; $U_{m2.e} = \exp(-0{,}115 \cdot 0{,}065 \cdot 80)(\cos(1.088) - j \sin(1088))/2 =$ $= 0{,}1285 + j0{,}243$; $U_{m0.e} = \exp(-0.115 \cdot 0.95 \cdot 80)(\cos 13.736 - j \sin 13.736)/2 \approx 0$	$U_{m1.e} = -\exp(-0{,}115 \cdot 0.033 \cdot 80)/3 = -0.246$; $U_{m2.e} = 0$; $U_{m0.e} \approx 0$
400	$U_{m1.e} = \exp(-0{,}115 \cdot 0{,}069 \cdot 80)/6 = 0{,}0884$; $U_{m2.e} = \exp(-0{,}115 \cdot 0{,}22 \cdot 80)(\cos(2976) - j \sin(2976))/2=$ $0{,}065 + j0{,}0109$; $U_{m0.e} \approx 0$	$U_{m1.e} = -\exp(-0{,}115 \cdot 0{,}069 \cdot 80)/3 = -0{,}177$; $U_{m2.e} = 0$; $U_{m0.e} \approx 0$
800	$U_{m1.e} = \exp(-0{,}115 \cdot 0{,}1 \cdot 80)/6 = 0{,}0664$; $U_{m2.e} = \exp(-0{,}115 \cdot 0{,}39 \cdot 80)(\cos(4{,}56) - j \sin(4{,}56))/2=$ $0{,}002165 + j0{,}0137$; $U_{m0.e} \approx 0$	$U_{m1.e} = -\exp(-0{,}115 \cdot 0{,}1 \cdot 80)/3 = -0{,}1328$; $U_{m2.e} = 0$; $U_{m0.e} \approx 0$

Table 4.3 Phase voltages at the end of the line with different transmitter coupling schemes at the beginning of the line

Frequency, kHz	Phase voltages at the side of the line at signal source connection	
	To phase 1	To phase 2
100	$U_{ph1.e} \approx 0{,}123 + 0{,}1275 - j0{,}243 = 0{,}341 - J\,0{,}243$; $U_{ph2.e} \approx -2{\cdot}0{,}123 = -0{,}246$; $U_{ph3.e} \approx 0{,}123 – 0{,}1275 + j0{,}243 = -0{,}0045 + j0{,}243$	$U_{ph1.e} \approx -0{,}246$;$U$ph2.e $\approx 0{,}492$; $U_{ph3.e} \approx -0{,}246$
400	$U_{ph1.e} \approx 0{,}088 + 0{,}065 - j0{,}0109 = 0{,}023 - j0{,}0109$; $U_{ph2.e} \approx -2{\cdot}0{,}088 = -0{,}1746$; $U_{ph3.e} \approx 0{,}088 + 0{,}065 + j0{,}0109 = 0{,}153 + j0{,}0109 = 0{,}153$	$U_{ph1.e} \approx -0{,}177$; Uph2.e $\approx 0{,}354$; $U_{ph3.e} \approx -0{,}177$
800	$U_{ph1.e} \approx 0{,}0664 – 0.00217 + j0{,}0137 = 0{,}0642 + j0{,}0137$; $U_{ph2.e} \approx -2{\cdot}0.0664 = 0.1336$; $U_{ph3.e} \approx 0{,}0664 + 0.00217 - j0{,}0137 = 0{,}0686 + j0{,}0137 = 0.153$	$U_{ph1.e} \approx -0{,}133$; $U_{ph2.e} \approx 0{,}266$; $U_{ph3.e} \approx -0{,}133$

Based on the received voltage values at the end of the line, it is possible to estimate attenuation of the line path when connecting the transmitter using phase 1-to-earth or phase 2-to-earth coupling scheme (see Fig. 4.2) and receiver according to one of the following schemes: phase r -to-earth (r = 1 outer phase; r = 2 center phase; and r = 3 other outer phase). The calculation was performed using the following formula:

$$a = 20\log\left|\frac{1}{U_{\text{ph}r.\text{e}}}\right| \tag{4.7}$$

Results of calculation according to (4.7) are shown in Table 4.4.

It can be seen from comparison of analysis results and the calculation using the software that they are almost identical, which is to be expected.

The structure of calculation formulas at each step of the line path attenuation calculation method based on the modal theory, attenuation for various paths based on the results of the calculation during analysis, and results of calculation based on these formulas allow explaining the reasons determining such a different nature of the line path frequency response for different coupling schemes.

It can be seen from these formulas that:

1. When coupling to a 2-0/2-0 line (center phase-to-earth), only one (first) mode is excited at the beginning of the line from the interphase modes. In this case, attenuation of the line path monotonously increases with increasing frequency in accordance with the change of the attenuation constant at changing frequency of the first mode. Such coupling schemes are called single-mode. They are the best and recommended for use.
2. When 1-0/1-0 scheme is used to connect to the line (outer phase-to-earth), two interphase modes are generated at the beginning of the line. The voltage vector of the coupling phase at the receiving end is determined by a linear combination of modal component voltage vectors at the same point.

Table 4.4 Line attenuation for different coupling schemes of HF paths

	Attenuation, dB, for the frequency, kHz		
Coupling scheme	100	400	800
1-0/1-0 (3-0/3-0)	9.14/9.2	31.9/31.4	23.6/22.5
1-0/2-0 (3-0/2-0)	12.18/12.3	15.1/15.13	17.5/17.7
1-0/3-0 (3-0/1-0)	12.35/12.2	16.3/16.3	23.1/232
2-0/1-0 (2-0/3-0)	12.2/12.3	15.0/15.1	17.5/17.7
2-0/2-0	6.2/6.3	9.0/9.3	11.5/11.8

Note. (1) Designation of 1-0/2-0 type means coupling the source to phase 1 and the receiver to phase 2; (2) in the attenuation columns, the numerator is calculated in the analysis, and the denominator is calculated using the WinTrakt software; (3) due to the symmetry of the outer phases' location relative to the center phase, the results obtained are also valid for coupling to phase 3 (another outer phase)

We have already mentioned that the attenuation constants and, most important, propagation velocities of the modes are different.

This leads to the fact that at the receiving end of the line, voltage vectors of these modes diverge by the angle $\Delta\beta_2 L$, and this value increases with increasing frequency in accordance with the formula (4.6).

At the same time, at a frequency for which $\Delta\beta_2 L = 3.14$ rad (180°), the coupling phase voltage is defined by the voltage difference of the first and second modes and can have a very small value) and the line path attenuation – a very large value. This frequency is called the attenuation pole frequency.

In the example above, the attenuation pole frequency is more than 400 kHz. At this frequency, the phase 1 voltage at the receiving end of the line is (in relative units, r.u.) approximately 0.023 r.u. with the voltage module for the first mode 0.088 r.u. and for the second mode 0.065 r.u.

Such coupling schemes are called multimode, and they are recommended for use with limitations imposed on maximum frequency of the frequency range used.

It should be noted that, if, while leaving the transmitter connected to phase 1, to connect the receiver to a center or second outer phase, the path attenuation in the area of the attenuation pole significantly reduces as only first mode voltage, with module equal to 0.175 r.u., present at phase 2 (center), but at phase 3, at the same voltages of the first and second mode as at phase 1, components of linear combinations, unlike phase 1, will be added but not subtracted, and the phase 3 voltage will be equal to the sum $0.088 + 0.065 = 0.153$ r.u. (instead of 0.023 r.u. for the first phase).

So, the curve of the frequency response of the line path of non-transposed overhead lines for different schemes for coupling to the line shown in Fig. 4.5 depends on the number of phase-to-phase modes involved in transmitting the signal from the source to the receiver.

For the single-mode coupling scheme, the frequency response characteristic increases monotonously.

For multimode schemes, visible attenuation poles can appear in the line path frequency response characteristics at the attenuation pole frequencies. The degree of attenuation increase in the attenuation pole depends on the ratio of components in the linear combination of the mode voltage vector involved in the signal transmission. The more the difference is between modules of mode linear combination voltage components, the less the attenuation pole. Since the attenuation pole is not caused by a "resonant" increase in a mode loss but by unfavorable phase relations between the mode voltages in the coupling phase, connecting the receiver to another phase for a signal frequency equal to the frequency of the attenuation pole will always lead to a decrease in the line path attenuation. This is the basis for application of "cross" coupling schemes, when the transmitter and receiver are connected to different phases.

Features of the line path frequency response for different schemes for coupling to non-transposed three-phase overhead lines with a horizontal phase arrangement were considered above. These features due to the multimode nature of signal propagation are typical for overhead lines with a different type of phase conductor and groundwire arrangements (triangular arrangement in a single-circuit overhead line, double-circuit overhead line) and for underground cable power transmission lines. At the same time, it should be noted that the single-mode coupling scheme in its pure form can only be implemented in the overhead line with a horizontal phase arrangement and in cable lines. In OPL with different phase arrangements, only "quasi-single-mode" schemes can exist in which, in addition to minimal loss mode, other modes with large losses are excited, but the voltage of modes with a large attenuation in the beginning of the line is much less than the voltage of mode with lower losses, and, therefore, increased attenuation in the area of the attenuation pole can be neglected.

As was said at the beginning, all the above arguments were related to non- transposed overhead lines. In principle, there are no single-mode (and "quasi-single-mode") coupling schemes for transposed OPL, since each mode wave approaching to the transposition point excites waves in the line after the transposition, in general, in all n modes. This makes the propagation process multimode, even if only one mode was generated at the beginning of the line. Thus, the attenuation poles are present in any transposed overhead line coupling scheme. At that, the attenuation pole frequency depends on the transposition scheme and selected coupling scheme. Coupling schemes in which the attenuation pole is in a higher-frequency range are called optimal. They should be preferably used for channel creation. Schemes in which the attenuation pole lies in the lower frequency range are called suboptimal.

For overhead lines with the standard transposition scheme shown in Fig. 4.6, when phases are horizontally arranged, optimal schemes are those with coupling to the phases that either starts or ends as the center one. In Fig. 4.6, these are phase A-to-earth, phase B-to-earth, and phase A-to-phase B.

For transposed overhead lines, the increase in the line path attenuation in the attenuation pole is usually less than for non-transposed overhead lines. This is because, at each transposition point, as already noted above there is an energy exchange between different modes, complicating conditions for propagation along the line in comparison with the case of a non- transposed overhead line. In this case, the probability of such unfavorable amplitude–phase ratios of different mode voltage vectors at the end of the line that leads to the appearance of the attenuation pole becomes very small.

To illustrate this information, results of calculation of attenuation for a line path with different schemes for connecting to transposed 220 kV overhead lines are presented in Fig. 4.7. For comparability of these results with the results obtained for the non-transposed overhead line, it is assumed that structure of the towers and phase conductors is the same as before. The length of the overhead line is assumed to be

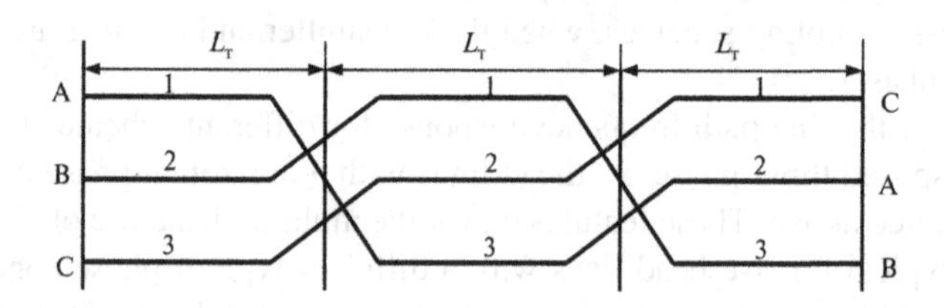

Fig. 4.6 Standard OPL transposition scheme

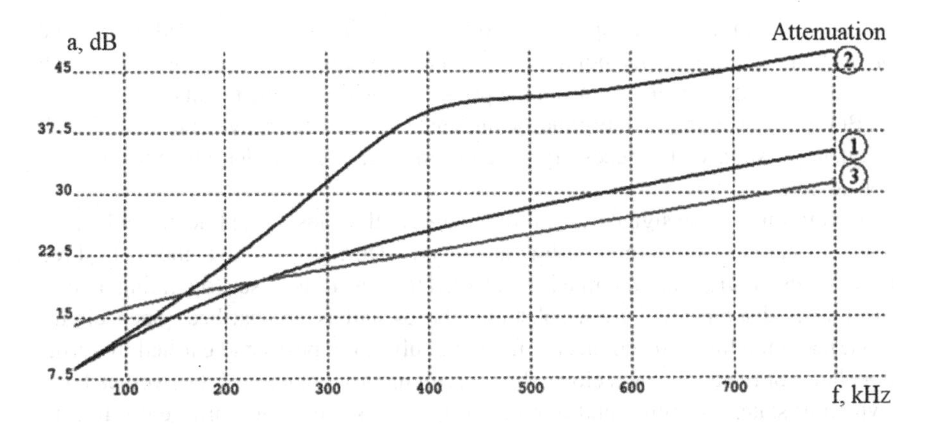

Fig. 4.7 Line path frequency response for various schemes of phase-to-earth coupling to transposed OPL 220 kV. *Graph 1* phase A (B)-to-earth, *graph 2* phase C-to-earth, *graph 3* phase A-to-phase B

180 km (transposition step is 60 km); the transposition scheme is standard. It is shown in Fig. 4.6.

The figure shows that for a given line with its transposition scheme and transposition step lengths:

1. For the non-optimal coupling scheme phase C-to-earth, the attenuation pole with an implicit increase in attenuation is located near the frequency of 420 kHz that corresponds to theoretical predictions of possible attenuation pole locations at $\Delta\beta_2 L_t \approx 112{,}5^\circ$ and 247.5°, where L_t is the length of the transposition step.
2. For optimal coupling schemes phase A-to-earth and phase B-to-earth, in the line path frequency response, the attenuation pole, theoretically located at a frequency at which $\Delta\beta_2 L_t \approx 360^\circ$, is not seen at all.

4.1.3 *Effect of Reflected Waves*

Earlier, we considered influence on the line path frequency response with a multi-mode nature of HF signal propagation under assumption that voltage waves reflected from non-homogeneity points of this overhead line can be neglected in the beginning of the considered OPL. This condition is usually satisfied in paths run along the long 330 to 750 kV lines which usually do not have tap lines or HF bridges.

In 35, 110 kV overhead line paths and often in 220 kV overhead line paths, this condition is usually not met. Let us consider the influence of reflected waves for the simplest case, when the path includes only one non-transposed overhead line and reflections occur only at the ends of the line. If there are more than two points where the line is not homogeneous (a path with a tap line or a HF bridge), reflections occur at each point where line homogeneity is distorted, and the line path frequency response becomes more complex. However, the result of analyzing the frequency characteristics for this example can also be applied in the case of a path with three or more non-homogeneity points, if we separately consider reflections from each pair of non-homogeneity points and apply superposition of the results.

Before proceeding to the issue of influence of reflected waves on the HF path frequency response, it is necessary to determine conditions under which this will be considered.

Conditions for the signal transmitting between the ends of the line depend on the load of phases of the overhead line at its ends. During operation, there are three basic states of OPL end: the line is connected to substation buses (operating mode), the line is disconnected from substation buses and is not earthed (open circuit mode), and the line is disconnected from the substation buses and earthed by earthing blades of linear disconnectors (short circuit mode). For each of the overhead line switching states, coupling and non-coupling phases are loaded differently for HF signal. Possible variants of phase conductor load impedance at the end of the overhead line are shown in Table 4.5.

Transition from one switching state of the overhead line to another state leads to the load change at the ends of the overhead line and, accordingly, to change to some extent the frequency response of the HF path attenuation. These changes will take place not only when the overhead line used for PLC channel is switched but also when other overhead lines which fit for the buses of the considered voltage of the end substations (due to changes in substation input impedance $Z_{\text{input.SS}}$) are switched.

Further, the influence of repeatedly reflected waves on frequency response is considered for two extreme load values of coupling and non-coupling phases – short circuit mode and open circuit mode.

The influence of reflected waves on the HF path frequency response in this case is manifested in a periodic increase and decrease in attenuation in the path relative to its value determined without considering influence of reflected waves. The frequency interval Δf, kHz, between the adjacent maximum and minimum attenuation is approximated by the formula:

$$\Delta f_{\max-\min} \approx \frac{300}{4L} = \frac{75}{L} \tag{4.8}$$

Table 4.5 Phase load conditions at the ends of the overhead line for different switching states of the line

OPL mode	Phase load impedance at the ends of the overhead line for: Coupling phase	Non-coupling phases
OPL is connected to substation buses (working mode)	Z_{CD} in parallel with $(Z_{LT} + Z_{\text{input.SS}})$	$Z_{\text{input.SS}}$
OPL is disconnected from substation buses (open circuit mode)	Z_{CD}	Isolated
OPL is disconnected from substation buses and earthed (short circuit mode)	Z_{CD} in parallel with Z_{LT}	Earthed

where 300 thousand km/s is the velocity of signal propagation in the lossless overhead line and L is the line length, km.

Let us consider the HF path frequency response defined considering reflected waves, as an example, for HF path which includes one non-transposed OPL, coupling devices, and line traps at the ends of the path. To reach continuity of the examples considered in different sections of this chapter, let us consider frequency characteristics for the same 220 kV overhead line with a horizontal phase arrangement, 80 km long, which was used for calculation in the examples of the Sect. 4.1.1.

Figure 4.8 shows the calculating HF path frequency response for the single-mode coupling scheme phase 2 (center)-to-earth and for the multimode coupling scheme phase 1 (outer)-to-earth. The frequency characteristics are given for two modes of the line switching state – open and short circuit.

Figure 4.8 shows that:

1. The frequency interval between adjacent maxima and minima attenuation (approximately 0.9 kHz) corresponds to interval calculated by (4.8).
2. The form of the HF path frequency response and attenuation non-uniformity depend on:

   – Switching state of the overhead line
   – Type of coupling scheme (single-mode/multimode scheme)

Let us look at the latest statements in more detail.

As mentioned above, the switching state of the overhead line determines the load of the coupling and non-coupling phases. This load, in turn, determines both transmitted signal power distributions between the modes at the beginning of the overhead line and coefficients of interphase wave reflection from the ends of the overhead line. They both affect the signal transmission conditions and, consequently, frequency characteristics of the attenuation.

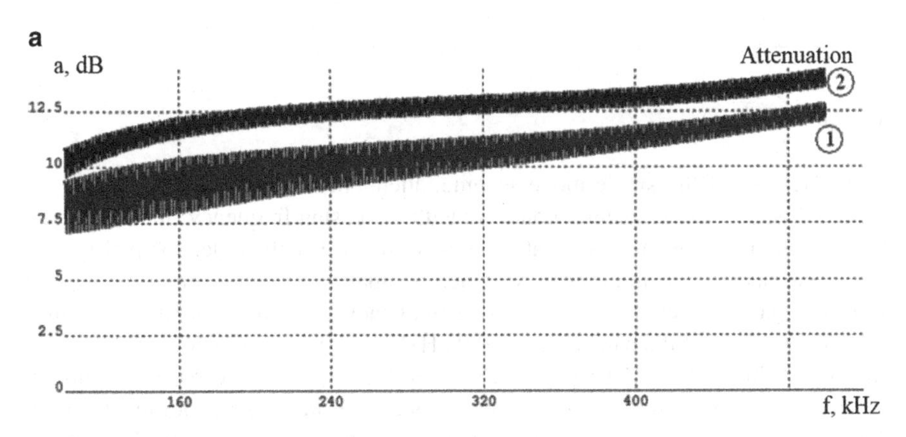

Fig. 4.8 Frequency response for the HF path influenced by reflected waves: (**a**) and (**b**) – center phase-to-earth coupling scheme (respectively, curve in wide frequency range and part of this curve in narrow range 105 to 120 kHz); (**c**) and (**d**) are the same for the outer phase-to-earth coupling scheme. *Graphs 1* for open circuit mode, *graphs 2* for short circuit mode

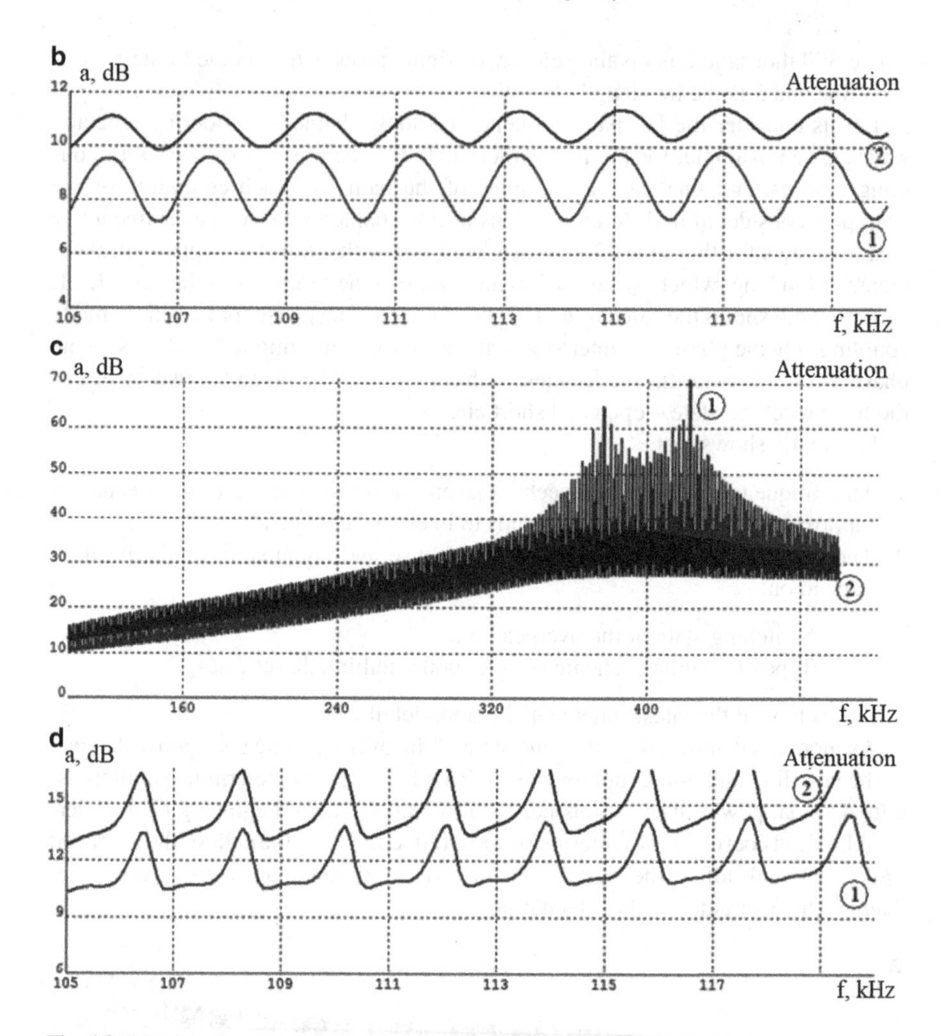

Fig. 4.8 (continued)

In the case of the single-mode scheme, attenuation non-uniformity caused by reflected waves decreases monotonously with increasing frequency (in accordance with the monotonous increase in attenuation of mode 1 in the reflection path).

In the case of the multimode scheme, attenuation non-uniformity caused by reflected waves increases with increasing frequency, reaching a maximum (in this case, up to 25 dB in the band of about 0.9 kHz) in the attenuation pole. It should be noted that this feature of frequency characteristics is confirmed by the results of measurements in existing overhead lines. This circumstance is an additional reason for introducing restrictions on application of multimode non-optimal coupling schemes.

Another feature of the HF path frequency response observed during switching should be noted. If, prior to switching, the OPL mode was open circuit on both OPL

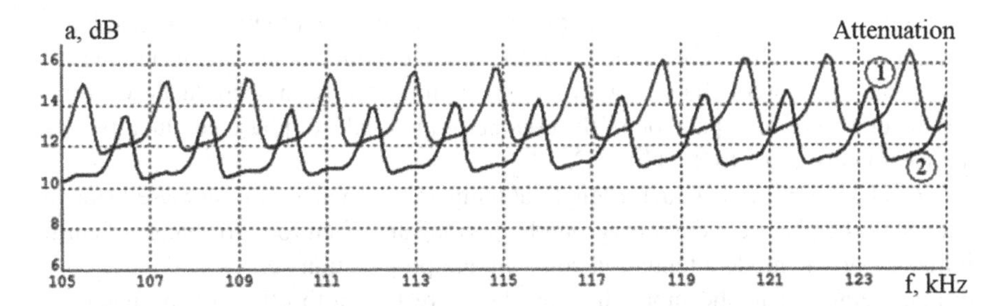

Fig. 4.9 HF path frequency response for outer phase-to-earth coupling scheme. *Graph 1* for open circuit mode on both sides, *graph 2* for open circuit mode on one side and short circuit on the other side of the OPL

sides and after switching became short circuit mode on both OPL sides, frequencies corresponding to maxima and minima of attenuation caused by reflected waves coincide for both modes. This is clearly seen in Fig. 4.8.

If, prior to switching, the OPL mode was open circuit on both sides and after switching became open circuit mode on one side and short circuit mode on the other side, frequencies corresponding to maxima of one mode will coincide with the frequencies of minima for other mode and vice versa. This is clearly seen in Fig. 4.9 which shows the HF path frequency response for this case.

4.1.4 *Effect of Weather Conditions*

Among the weather condition factors which affect attenuation of the HF path, let us consider air temperature and ice coating: ice hoarfrost and rime deposits (IHRD) on the overhead line conductors.

Let us consider the influence of air temperature on characteristics of the PLC channel. Phase conductors and cables in the span between two adjacent towers sag. This sagging is described by the chain line equation, which can be represented by a parabola for the overhead line with its ratio of span length and sag value. Due to sagging, the length of the phase conductors and groundwires in the span is longer than the span length. An idea of the difference between length of phase conductor and span length (ΔL_{ph}) can be obtained using the approximate formula:

$$\Delta L_{ph} = \frac{8 l_{sag}^2}{3 L_{span}}$$

where ΔL_{ph} is the difference between the length of the phase conductor and span length (L_{span}) and l_{sag} is the phase conductor sag.

It should be noted that the difference between the length of the phase conductor in the span and length of the span itself is small. So, at a span length of 400 m and a sag of 15 m, the difference between the length of the sagging phase conductor and span length is only 1.5 m.

Temperature of the phase conductor which depends on air temperature and loading current of the line affects length of the phase conductors and the groundwires. The conductor length changes with its temperature change in accordance with the coefficient of linear expansion of the conductor material (for phase conductors, it is approximately equal to $2 \cdot 10^{-5}$ m/°C.

The change of the conductor length at temperature increase or decrease leads to change of the phase conductors (groundwires) sag and, therefore, to change of earth influence on the mode attenuation constant and propagation velocity (at other conditions being equal, the more the sag, the more the earth effect on propagation constant).

In this case, the influence of temperature (and correspondent change of the conductor sag) on earth losses is different for different schemes for coupling to the line.

For single-mode coupling schemes, the earth influence on the main mode (usually mode 1) attenuation constant and the propagation velocity is very small. Therefore, changing the sag practically does not change attenuation of the path.

For multimode coupling schemes, the path attenuation at temperature change can be rather large. This change is due to both change in the mode attenuation constant (in this case, the first and second modes) and change in propagation velocity. At that, it is not the velocity change itself which is important but change in the difference between them for interphase modes. The less the sag, the less the difference and vice versa. Temperature change and, as a result, change of velocity difference in accordance with the formula (4.6) cause change of the attenuation pole frequency causing all consequences.

To illustrate this, attenuation versus frequency for HF path using single-mode and multimode coupling schemes for 220 kV overhead line described in Sect. 4.1.1 is shown in Fig. 4.10.

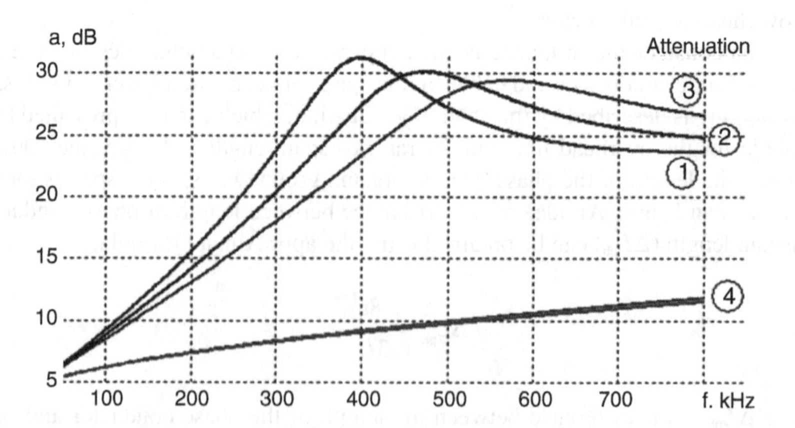

Fig. 4.10 Frequency response of the HF paths which use different coupling schemes and have different sags (15.0, 14.17, and 13.3 m): outer phase-to-earth – upper graphs (graph 1, 15 m sag; graph 2, 14.17 m; graph 3, 13.3 m); center phase-to-earth – lower graphs (4), almost identical for all sag values

Calculations are made for the phase conductor sag of 15 m (corresponding to a temperature of +40 °C) and 14.17 and 13.3 m (corresponding to a temperature of approximately +20 and 0 °C).

The calculation results given in this figure show that the stability of parameters of a single-mode coupling scheme when the air temperature changes should also be added to this scheme advantages.

Ice hoarfrost and rime deposits (IHRD) on overhead lines cause HF signal energy losses which occur in these deposit layers. These losses lead to change of modal parameters of the overhead line section with IHRD and, consequently, to change of parameters (including attenuation) of HF path in which part of the OPL conductor length being a part of the path is covered with IHRD.

IHRD on overhead line conductors can have different structures. The most general ideas about types of IHRD, their density, and conditions of occurrence, necessary for calculating HF path attenuation taking into account ice hoarfrost and rime, are as follows:

- Pure ice: density – 0.9 g/cm^3; temperature – about −3 °C
- Ice with air bubbles: density – 0.75 g/cm^3; temperature – about −3 °C
- Rime ice: density – 0.1 to 0.3 g/cm^3; temperature – −3 to −10 °C
- Hoarfrost: density – 0.05 g/cm^3; temperature −10 to −40 °C
- Ice snow mixture: density – 0.2 to 0.4 g/cm^3; temperature – 0 to −20 °C
- Wet snow: density – 0.2 g/cm^3; temperature – about 0 °C

Data on IHRD structure and the thickness of ice layer are obtained based on long-term observations conducted by hydrometeorological stations located in the area of the overhead line route.

As an example, let us describe main principles accepted in [12] for calculation of IHRD effect. Usually when designing PLC channels, the attenuation caused by IHRD is calculated for IHRD in the form of ice with a density of 0.9 g/cm^3 and temperature of −3 °C. The ice layer thickness is taken in accordance with the country territory zoning standard, for ice layer thickness repeatability one time every 5 years. The length of overhead line conductors covered with ice coating is assumed to be 30 km.

The experience of operating PLC channels using overhead lines which were designed considering normalized ice layer thickness with repeatability of one time in 5 years in most cases showed adequate reliability of these channels in ice coating conditions.

Therefore, additional attenuation caused by ice is determined in [12] in accordance with the climate zoning maps of the territory of the country for ice layer thickness stipulated in climatic maps for design of electric power transmission lines. Standard ice layer thickness (when reducing ice layer to a cylindrical shape and density of 0.9 g/cm^3) is given in Table 4.6.

Table 4.6 Ice layer thickness used for different icing areas, repeatability of one time in 5 years

Area category for icing in accordance with [12]	I	II	III	IV	Special
Standard ice layer thickness, mm, with repeatability 1 time in 5 years	5	5	10	15	20 and more

As mentioned above, IHRD on the overhead line section conductors leads to change of modal parameters of this overhead line section and, accordingly, to change of conditions for signal propagation along the overhead line which includes the considered section with IHRD. As an example, let us consider IHRD influence on modal parameters of a 220 kV overhead line with such a tower and phase conductor design which was already considered in Sect. 4.1.1 when modal parameters of a line without IHRD were analyzed.

In this case, data given in Sect. 4.1.1 will be used to get information on modal parameters of the overhead line when IHRD appears on the conductors. Let us assume that IHRD is in the form of pure ice with a layer thickness of 1 cm.

Analysis of the considered overhead line modal parameters, with phase conductors ice coating, shows that IHRD on the conductors:

- Practically does not affect the value of matrix elements λ and δ. The difference between the values of these matrix elements and their values in the absence of IHRD is negligible.
- Has very little effect on modal wave impedance. For example, when the ice layer thickness is 1 cm, the impedance modulus decreases by about 10%, and the wave impedance remains almost purely real and does not depend on frequency.

IHRD mainly influences attenuation constant α and phase constant β (propagation velocity υ) of different modes.

In Table 4.7, values of modal parameters of 220 kV overhead line mentioned above, for five frequencies, determined taking into account ice (layer thickness 1 cm, density 0.9 g/cm^3) are given. For comparison, the table also repeats data of Table 4.1 for values of the considered modal parameters of OPL without IHRD.

As can be seen from Table 4.7, energy losses in ice formations lead to approximately the same increase of the attenuation constant for each mode and such a decrease of the propagation velocity for each interphase mode which "equalizes" these velocities, making them very close in magnitude.

At all other conditions being equal, the degree of change of attenuation constant α and propagation velocity υ for different modes depends on IHRD thickness, its structure, conductor diameter, and whether the phase of the line is bundled or not.

Now, let us see how the HF path attenuation changes when ice (in the form of pure ice) appears on the conductors of the part of the overhead line length being a part of the HF path. Let us consider this for the example of the HF path for non-transposed and transposed 220 kV overhead lines which were considered in Sect. 4.1.2 (design of the overhead line is the same as for the line for which parameters are given in Table 4.7).

Non-transposed OPL. Figure 4.11 shows the frequency response of the HF paths according to the single-mode coupling scheme (phase 2-to-earth) and according to the multimode coupling scheme (phase 1-to-earth). Performing calculations using the WinTrakt software

Of 80 km of the overhead line length entering the path, 30 km is covered by ice with a layer thickness of 1 cm, and there is no ice on the 50 km section (in accordance with the [12] recommendations on frequency selection). In the same figure,

Table 4.7 Attenuation constant α, phase constant β, and propagation velocity υ for modes of 220 kHz OPL under consideration

Frequency, kHz	IHRD factor	Conditional mode name					
		Mode 1 center phase – outer phases		Mode 2 phase – outer phase		Mode 0 (Earth mode)	
		α	β/υ	α	β/υ	α	β/υ
50	without ice	0.023	1.0501/299.17	0.035	1.0578/296.99	0.56	1.162/270.36
100	with ice	0.064	1.1034/284.7	0.071	1.1047/284.4	0.60	1.1971/262.43
200	without ice	0.033	2.0987/299.38	0.065	2.1123/297.46	0.95	2.2704/276.74
400	with ice	0.167	2.1995/285.7	0.184	2.2011/285.5	1.07	2.3353/269.1
800	without ice	0.047	4.1951/299.54	0.12	4.2183/297.9	1.57	4.452/282.26
	with ice	0.373	4.3738/287.3	0.41	4.3755/287.2	1.83	4.565/275.3
	without ice	0.069	8.387/299.7	0.22	8.4242/298.3	2.52	8.7632/286.8
	with ice	0.576	8.7008/288.9	0.67	8.7003/288.9	2.9	8.9595/280.5
	without ice	0.1	16.769/299.75	0.39	16.826/298.74	3.92	17.312/290.35
	with ice	0.693	17.357/289.6	0.914	17.343/289.8	4.37	17.677/284.4

Notes: units of matrix elements: α dB/km, β rad/km, υ thousand km/s

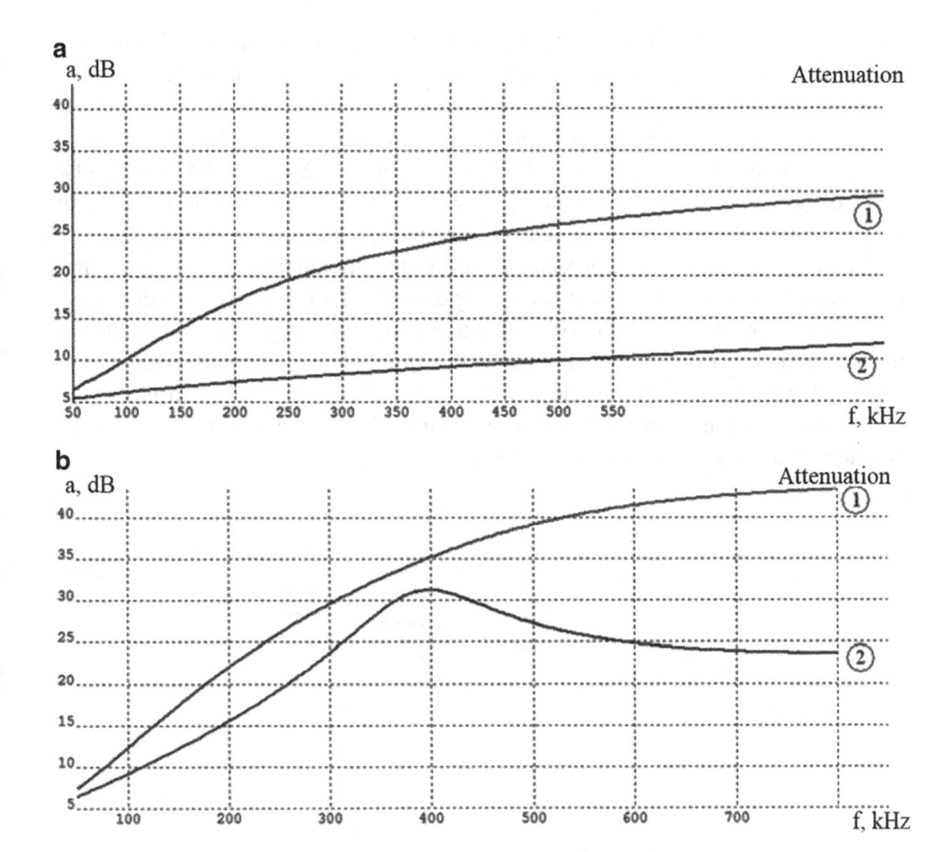

Fig. 4.11 HF path frequency response for various schemes of phase-to-earth coupling to 220 kV overhead line: (**a**) – center phase (2)-to-earth; (**b**) – outer phase (1)-to-earth. *Graphs 1* with ice, *graphs 2* without ice

for comparison, dependences of the HF path attenuation are also shown but without IHRD on the OPL (repeated data of Fig. 4.5). As can be seen from the figure, frequency dependence of additional attenuation caused by ice differs significantly for single-mode and multimode coupling schemes.

Figure 4.12 shows frequency response of additional attenuation caused by ice obtained as the difference of the HF path attenuation for corresponding coupling schemes with ice and without ice. It can be seen from Fig. 4.12 that, for the single-mode coupling scheme for connection to the overhead line, additional attenuation caused by ice monotonously increases with increasing frequency.

From the analysis of HF path which uses the single-mode coupling scheme described in Sect. 4.1.2, the attenuation of the HF path (regardless of ice taking into account) is defined by the attenuation constant of the corresponding interphase mode. At that, increase of the HF path attenuation at ice presence (Δa_{IHRD}) can be defined as follows:

$$\Delta a_{\mathrm{IHRD}} = \Delta \alpha_{\mathrm{IHRD}} L_{\mathrm{IHRD}}$$

where $\Delta \alpha_{\mathrm{IHRD}}$ is attenuation constant increase for considered mode, dB/km (this value does not differ very much for different modes) and L_{IHRD} is length of the overhead line section with ice.

So, for example, according to Table 4.7, at a frequency of 400 kHz, attenuation constant increase for the first mode due to ice is equal to $\Delta \alpha_{\mathrm{IHRD}} = 0.576 - 0.069 = 0.507$ dB/km that at the length of L_{IHRD} = 30 km gives attenuation increase of Δa_{IHRD} = 15.2 dB. The same value can be seen in Fig. 4.12.

For the multimode coupling scheme, additional attenuation caused by ice has quite a complex behavior (in contrast to single-mode). As it follows from the analysis conducted in Sect. 4.1.2, the HF path attenuation (and frequency dependence of this attenuation) for multimode schemes is determined not only by attenuation constants of the modes involved in the signal propagation but also by the difference of these modes' propagation velocities. It is obvious that the same can be said about the attenuation increase due to ice for these schemes.

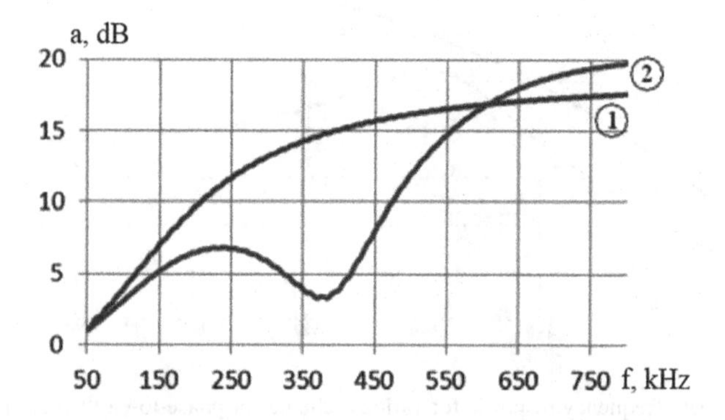

Fig. 4.12 Frequency response of additional attenuation caused by ice for length of the line section with ice 30 km. *Graph 1* single-mode coupling scheme, *Graph 2* multimode coupling scheme

As can be seen from Table 4.7, ice changes not only attenuation constants of different modes, increasing them, but also reduces the propagation velocity of different mode waves, reducing difference between these velocities.

It is this last circumstance that determines behavior of frequency response of addition to attenuation due to ice for multimode schemes: the smallest increase in attenuation due to ice corresponds to frequency range close to the attenuation pole frequency for the HF path defined without ice. As we noted earlier, the attenuation pole frequency corresponds to the frequency at which the phase shift between voltage (current) vectors of interphase modes involved in signal propagation is equal to 180°. This shift is defined for the mode voltage vectors in the coupling phase at the receiving end of the overhead line.

For a path with ice coating, this shift is reduced due to the fact that it is significantly reduced in the part of the overhead line (in the section with ice and in this case, it is practically absent).

This circumstance leads to some decrease of HF path with ice attenuation due to the shift of the attenuation pole in the path with ice to the higher-frequency range (similar to influence of ambient air temperature considered above). Because of this, attenuation decrease due to ice is especially noticeable in the frequency range adjacent to the attenuation pole defined for ice-free overhead line (in this case, about 400 kHz). The more the attenuation rises at the attenuation pole frequency, the more significant is the decrease of the path with ice attenuation near this frequency.

It should be said that the analysis of frequency response behavior for attenuation increase due to ice in the frequency range near the attenuation pole of the path without ice for multimode coupling schemes is more theoretical than applied, since use of multimode coupling schemes to create channels, especially in the frequency range close to the attenuation pole, is not recommended for the reasons discussed in this chapter.

Transposed OPL. Figure 4.13 shows frequency dependences of attenuation of the path along transposed 220 kV overhead line with a length of 180 km parameters of which are given in Sect. 4.1.2. Frequency response characteristics are given for optimal (phase A-to-earth and phase B-to-earth) and non-optimal (phase C-to-earth) coupling schemes discussed above (see Fig. 4.7). Ice with a layer thickness of 1 cm was added on the line section with a length of 30 km, located at the end of the overhead line.

In the same figure, for comparison, dependences of the HF path attenuation without IHRD on the OPL are also shown (repeated curves of Fig. 4.7). The more the attenuation rises at the attenuation pole frequency, the more significant is the decrease of the path with ice attenuation near this frequency.

Figure 4.14 shows frequency dependences caused by ice additional attenuation obtained as the difference of the HF path attenuation for corresponding coupling schemes with ice and without ice obtained according to data of Fig. 4.13.

It can be seen from Fig. 4.14 and its comparison with Fig. 4.12 that, for optimal coupling scheme to the overhead line (phase A-to-earth or phase B-to-earth), the value of additional attenuation caused by ice practically coincides with the corresponding value for the single-mode scheme for connection to non-transposed overhead line.

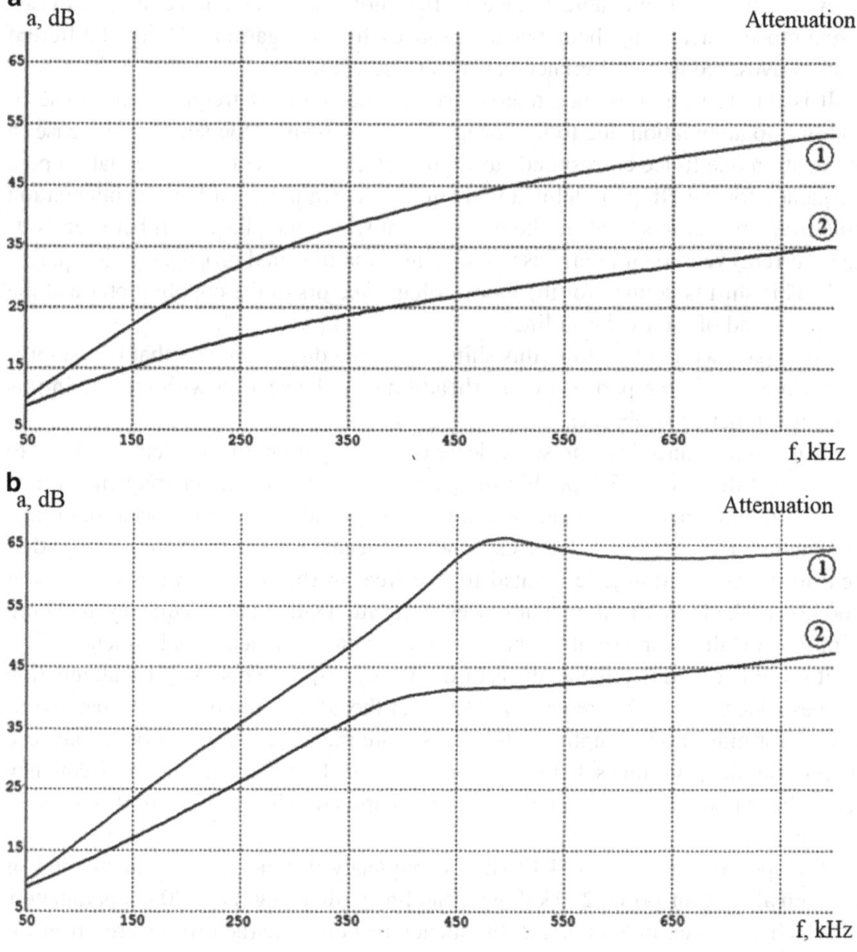

Fig. 4.13 HF path frequency response: (**a**) – optimal coupling scheme phase A-to-earth; (**b**) – non-optimal coupling scheme phase C-to-earth. *Graphs 1* with ice, *graphs 2* without ice

For a non-optimal coupling scheme (phase C-to-earth), the shown behavior of the frequency response of additional attenuation due to ice is caused by the phenomena analyzed for the multimode coupling scheme for connection to non-transposed overhead line. The only difference is that, as stated in Sect. 4.1.2 for non-optimal coupling scheme, the increase of attenuation near the attenuation pole frequency is usually expressed implicitly. Therefore, the phenomena which lead to a monotonous increase of the additional attenuation due to ice with increasing frequency are not so significant ("smoothed out"). This is clearly seen from comparison of graph 2 in Fig. 4.12 and graph 1 in Fig. 4.14.

Earlier, we considered additional attenuation introduced into the IHRD path in the form of pure ice with a density of 0.9 g/cm^3.

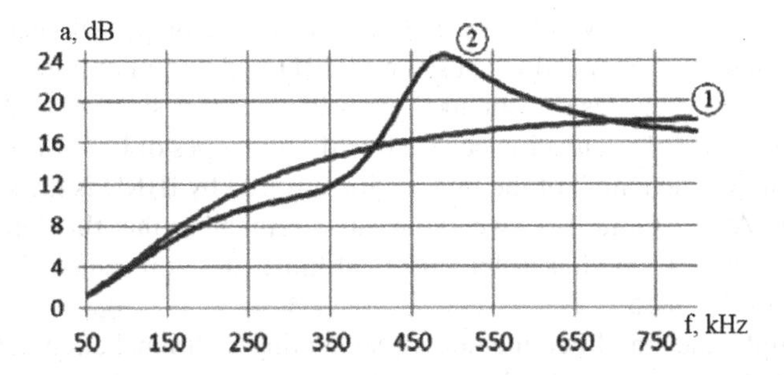

Fig. 4.14 Frequency response of additional attenuation caused by ice for the length of the line section with ice 30 km. *Graph 1* phase A-to-earth, *Graph 2* phase C-to-earth

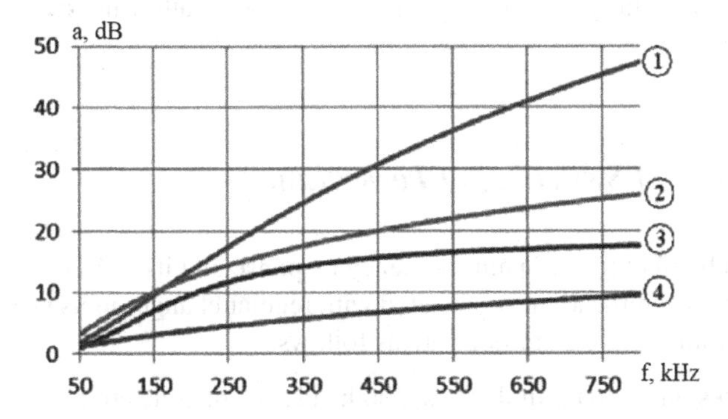

Fig. 4.15 Additional attenuation caused by IHRD of different forms into the path under consideration. *Graph 1* rime ice (0.2 g/cm³; −7 °C), *graph 2* rime ice (0.2 g/cm³; −20 °C), *graph 3* pure ice (0.9 g/cm³; −3 °C), *graph 4* hoarfrost (0.05 g/cm³; −20 °C)

Let us consider how other IHRD forms described above affect HF path attenuation. The simplest for comparison case of the single-mode scheme for connection to non-transposed overhead line will be considered. As an example, let us do this on an already considered 220 kV overhead line with a length of 80 km and the length of the overhead line section covered with IHRD of 30 km. The IHRD layer thickness is 1 cm. Such a representation of the length of the overhead line section with IHRD in the form of rime of hoarfrost, although not always consistent with the practice in part of the length of the overhead line section with IHRD, is convenient for comparison of the results obtained for different IHRD forms.

There is a common misconception that pure ice with a density of 0.9 g/cm³ causes larger losses (large attenuation) to the signal propagation path than rime ice which has a lower density.

Calculations for the considered path performed taking into account physical properties of various forms of ice in a wide frequency range are shown in Fig. 4.15.

As can be seen from this figure, the largest losses are caused by IHRD with a density of about 0.2 to 0.4 g/cm³ having the form of rime ice and wet snow mixture (at temperatures range up to −10 °C). This is especially noticeable in the frequency range above 200 kHz.

Hoarfrost with a density of 0.05 g/cm^3 causes attenuation per kilometer which is significantly less than attenuation caused by IHRD in the form of pure ice.

When analyzing the results obtained from Fig. 4.15, it should be remembered that the comparison of attenuation caused by different types of IHRD is given under the condition that the length of the line section covered by IHRD is the same for all IHRD types. At the same time, we should also remember that this length for ice deposits can exceed the chosen length of 30 km that significantly changes ratio obtained from Fig. 4.15.

For example, the simplified method of calculating additional attenuation caused by ice adopted in [12] is valid only for single-mode coupling schemes and for short overhead lines, when the theory of symmetric overhead lines can be used for calculation. In all other cases, it is recommended to use for calculation the WinTrakt software, which allows defining parameters of the path with ice without any limitations.

4.1.5 Effect of Switching of Power Lines

Factors which affect the HF path frequency response and its stability over time are switching of the overhead line associated with regular changes in its switching state during operation. These conditions are as follows:

- Line works under load. In this case, both sides of the line are connected to substations buses.
- Line works in hot standby mode. In this case, one end of the line is connected to the substation and energized, and at the other end of the line, the phases are isolated (the line is disconnected and not earthed).
- Line is disabled. In this case, the line is usually disconnected by linear disconnectors and earthed by earthing blades of linear disconnectors.

At that:

- When the line is working, input impedance of the end substation may change due to disconnection (connection) of other overhead lines of the same voltage connected to this substation.
- In the process of removing the overhead line from operation, significant time may pass between its disconnection and further earthing. In addition, the moments of its disconnection and the moments of its earthing on the end substations, as a rule, do not coincide in time.

When considering the effect of reflected waves on the HF path frequency response characteristics, we have already discussed this and referred to Table 4.5 with instructions on changing the load of operating and non-operating phases of the overhead line at high frequencies. We will continue this analysis here.

It can be assumed that the change of the OPL switching state and the corresponding attenuation change occur abruptly (transient processes at OPL transition from one switching state to another can be neglected).

Having in mind that the DPLC channel should not lose the performance when overhead line is switched or abrupt attenuation change caused by these switches occurs, in this analysis we should pay attention not only to the changing attenuation at a switching condition but also to the magnitude of the attenuation "jump" at transition from one switching state to another.

The change of the HF path frequency response when changing the overhead line switching state occurs not only in a HF path consisting of a single overhead line without tap line, as was discussed in Sect. 4.1.2, but also in any complex HF path with tap lines and HF bridges. As an example, let us consider a tap line in the HF path.

Features of frequency response of the HF path caused by a tap line which is not used for communication are determined by influence of waves reflected (with a reflection coefficient whose modulus is (1) from the end of the tap line. These waves lead to periodic change of the branch substation input impedance from minimum to maximum values and, as a result, to periodic change of the HF path attenuation caused by the tap line. Frequency interval between adjacent maximum and minimum of the input attenuation is defined from the approximate formula (4.8) in which length L is assumed to be equal to the length of the tap line.

To reduce this attenuation, line traps are included at the beginning of the tap line that limits change of the input impedance and attenuation caused by the branch substation.

As an example, we will show change of attenuation for the HF path with a branch substation when the switching state of the tap line at its end on the branch substation is changed (line is connected to the branch substation and disconnected from the branch substation with a possible earthing) for a 110 kV overhead line with a length of 40 km, with a tap line length of 1 km, and with a line trap in the point of the tap line connecting to the main line. The place where the tap line is connected is 15 and 25 km away from the end substations.

Figure 4.16 shows the frequency response of the HF path for two specified modes of the tap line at its end.

The figure clearly shows periodic change of attenuation due to two reasons:

1. The presence of three reflection points in the main overhead line (its ends and tap line connection point). Wave reflection between these three points leads to periodic change of the path attenuation with a frequency interval between adjacent maxima and minima of attenuation determined by the overhead line length (40 km) and its sections between the overhead line ends and tap line connection point (15 and 25 km). In accordance with (4.8), for reflection from different points, this interval is in the range 2–5 kHz;
2. Full wave reflection from the end of the tap line that leads to change of attenuation caused by the tap line. Frequency interval between the adjacent maximum and minimum attenuation approximately defined by (4.8), which is equal to 75 kHz for a tap line length of 1 km.

Stepwise attenuation change during switching of the tap line at its end from one state to another depends on the frequency range point in which this change is defined. In this example, it varies between 0 and 4 dB.

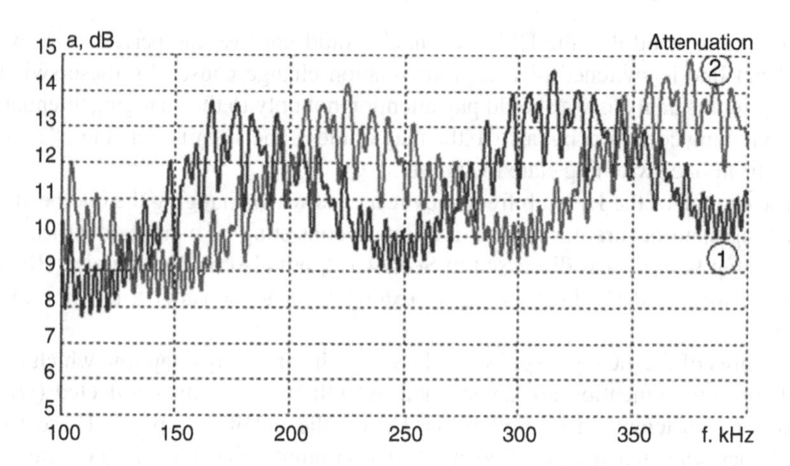

Fig. 4.16 Frequency response of the HF path for 110 kV overhead line with a tap line. *Graph 1* overhead line is connected at all three ends, *Graph 2* tap line is disconnected from the branch substation

4.1.6 Frequency Response of the HF Path with HF Bridge

Figure 4.17 shows a diagram of the HF path with HF bridge at the intermediate substation.

Let us describe features of frequency response of HF path with HF bridge. In some cases total attenuation of HF path between SS1 and SS3 (see Fig. 4.11) may differ significantly from the sum of HF path attenuations of its sections (between SS1 and SS2 and between SS2 and SS3). In this case, in one part of the bandwidth of the coupling device included in the bridge scheme at SS2, the path attenuation can be more than the sum of attenuation of the path sections and in the other part – less than this amount.

This phenomenon can be observed in a widespread class of HF bridge schemes in which:

- The HF path is formed with the phase-to-earth coupling scheme.
- OPL1 and OPL2 lines (see Fig. 4.17) which form the HF path are of the same voltage class.
- Only two lines with a voltage under consideration are connected to the intermediate substation with HF bridge along which the HF path with HF bridge is formed (SS2 in Fig. 4.17).
- The phase along which the HF path is formed (coupling phase) is the same both before the HF bridge (line 1) and after the HF bridge (line 2). Let us consider the mechanism of signal transmission through the intermediate SS2 for the signal transmission direction, for example, from SS1 through SS2 to SS3. Let us assume that both overhead lines (1 and 2) have a horizontal phase arrangement and the path is formed according to a single-mode scheme, center phase (2)-to-earth. Selection of this phase arrangement and coupling scheme makes the consideration more obvious. All analyzed phenomena will also occur in overhead lines with a different phase arrangement.

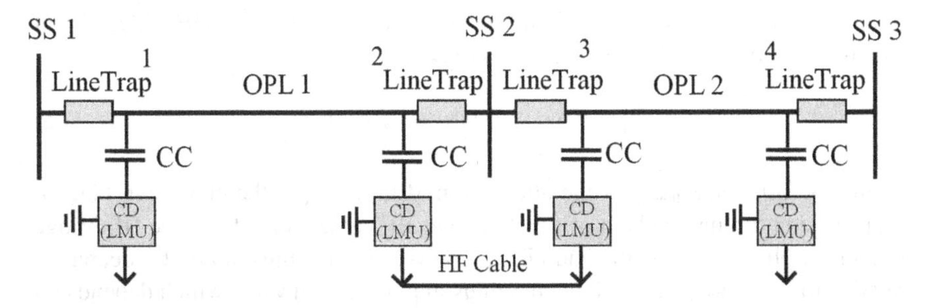

Fig. 4.17 HF path with HF bridge at the intermediate substation

Let us make some assumptions which will significantly simplify considering of the issue without compromising its essence:

- Let us assume that the line trap in the SS2 HF bridge scheme is ideal, i.e., it has an infinitely large blocking impedance over an entire frequency range under consideration.
- Let us neglect the reflection of the incident wave from the end of line 1. In this case, relative voltage vectors of the OPL1 phases at its end are equal to the voltage vectors caused by the only incident wave of the interphase mode in this point.

Considering the assumptions, the signal transmission process is as follows (see Fig. 4.17):

1. In point 1 (at SS1 at the beginning of OPL1), zero (earth) mode and 1st interphase mode are excited.
2. Between points 1 and 2 (along the OPL1 length), the voltage of the incident zero wave is almost completely fading, and only the incident wave of mode 1 comes to SS2 (with the corresponding attenuation).
3. In point 2 (at SS2 at the end of OPL1), relative voltage vectors of the OPL1 phases are equal to relative voltage vectors of the incident wave of interphase mode 1 at the point under consideration.

Between points 2 and 3 (signal transmission from OPL1 to OPL2 via SS2), transmission of phase voltage vectors occurs both through the coupling phase 2 (through an especially created HF bridge) and through the non-coupling phases 1 and 3 (which are connected to each other in OPL1 and OPL2 via busbars of substation 2). In this case, the SS buses are represented by a capacitor with a relatively small (about 5000 pF) capacity, and shunt action of the buses, in accordance with the assumptions, is neglected. This capacity is equal to the sum of the capacities which makes the high-voltage equipment connected to the busbars of the outdoor switchgear for the voltage in question equivalent (busbars themselves, switches, disconnectors, surge arrester, current and voltage measuring transformers, and power transformers). In this case, relative voltage vectors of the non-coupling phases 1 and 3 in point 3 (the beginning of OPL2) are equal to the voltage vectors of the same phases in point 2 (at the end of OPL1), and the relative voltage vector of the coupling phase 2 in point 3(U_{23}) is related with the voltage vector of the same phase in point 2 (U_{22} by the formula describing signal transmission through HF bridge circuit consisting of two coupling devices and HF cable:

$$U_{23} = U_{22b} e^{-0{,}115\alpha_{\text{HFbridge}}} \cdot e^{-j\beta_{\text{HFbridge}}} \quad (4.9)$$

In this formula, a_{HFbridge} is the attenuation, dB; β_{HFbridge} is the phase constant for transmission through a chain of two CDs and HF cable which form the HF bridge scheme at phase 2. Thus, the coupling phase voltage modulus in point 3 decreases relative to the same phase voltage modulus in point 2 by a value which depends on attenuation of a_{HFbridge}, i.e., approximately 1.1 to 1.6 times, and the shift angle between these voltage vectors changes with frequency within the nominal bandwidth of the CD by a value from 0 to 360°.

4. In point 3 (at SS2 at the beginning of OPL2), in phases 1, 2, and 3, the voltage system is decomposed into the incident wave voltage in zero and 1st interphase modal components which then propagate along OPL2 toward SS3. The proportion of voltage generated as zero and 1st interphase modes depends on the ratio between relative voltage vectors of phases 1, 2, and 3 in point 3.
5. Between points 3 and 4 (along the OPL2 length), zero-mode voltage of incident wave almost completely fades, and the incident wave comes to SS3 with the corresponding attenuation only as the first mode.
6. In point 4 (on SS3 at the end of OPL2), the coupling phase voltage (2) is defined by the incident wave voltage in the first interphase modal component.

The main factor determining behavior of the attenuation of the HF path under consideration is changing angle between voltage vectors of the coupling and non-coupling phases in point 3.

At frequencies where this angle is equal to 0° (angle β_{HFbridge} in (4.9) is equal to π radians or 180°), the proportion of signal power which falls into zero-mode wave is maximum and interphase mode – minimum. In this case, the signal reception level at phase 2 of SS3 is minimum possible, and, therefore, the attenuation of the path with a HF bridge is maximum possible. In a hypothetical case when the coupling and non-coupling phase voltage modules are equal, all the signal energy of the beginning of the OPL2 goes to zero mode, and the attenuation of the HF path will be defined by the attenuation of zero mode at the OPL2 length and will be very large.

In a certain frequency range, the attenuation of the HF path with the HF bridge will be more than the sum of attenuations of the path sections before and after the HF bridge.

At frequencies where this angle is 180° (angle β_{HFbridge} in (4.9) is 0 or 2π rad or 0 or 360°), the proportion of signal power which falls in zero-mode wave is minimum, and interphase mode is maximum. In this case, the signal reception level at phase 2 of SS3 is maximum possible, and, therefore, the attenuation of the path with a HF bridge is minimum possible. In a hypothetical case, when the coupling and non-coupling phase voltage modules' relation is 1:(−2):1, all the signal energy at the beginning of OPL2 falls into the interphase mode, and in a certain frequency band in the area of this frequency, the attenuation of the HF path with HF bridge will be less than the sum of attenuation of the path sections before and after the bridge.

Now, after we have analyzed the signal propagation process from the beginning to the end of the HF path, we will give the frequency response for the path with a HF bridge on a particular case. Let us consider the HF path shown in Fig. 4.17 with OPL1 and OPL2 lengths equal to 30 and 40 km, respectively. The coupling device (line matching unit) used in calculation has a capacity of 6400 pF and bandwidth of 36–140 kHz. The HF cable length is 15 m.

In Fig. 4.18, the frequency response for the HF path under consideration with a HF bridge within the CDs' bandwidth is given. In addition, the same characteristic defined for the path sections is shown for comparison.

The figure clearly shows areas of increased attenuation (at the edges of the frequency range under consideration) and reduced attenuation (in the center of the frequency range under consideration).

In the case, when nominal frequency bands of the PLC channel operating on the HF path are in the frequency range of increased attenuation, this attenuation can be reduced. To do this, as can be seen from (4.9), it is necessary to change the phase constant $\beta_{\mathrm{HFbridge}}$ by π (or 180°), thereby swapping location of the areas of increased and reduced attenuation.

To do this, conductors connected to the beginning and to the end of the "cable" windings of the air transformer of one of the coupling devices in HF bridge scheme can be mutually swapped.

This is clearly seen in Fig. 4.19 which shows attenuation for the considered path for variants without changing the phase in one of the coupling devices and with the phase change.

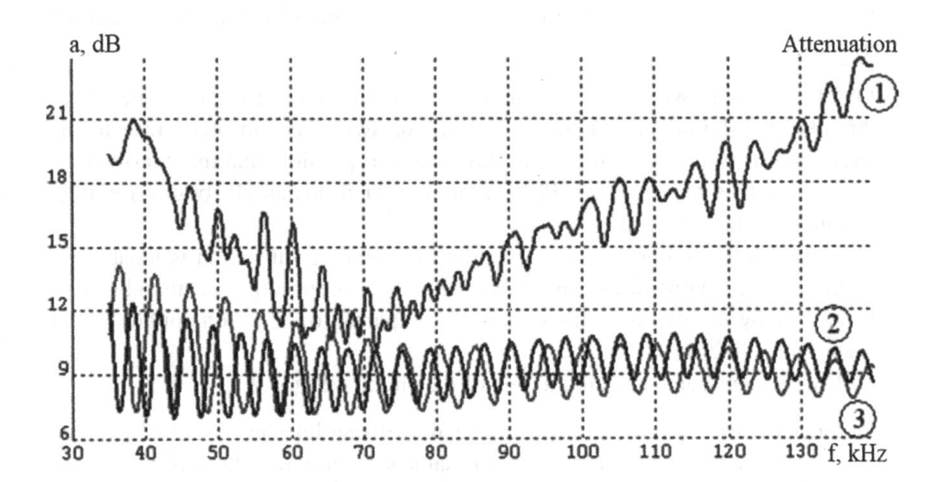

Fig. 4.18 Frequency response for the HF path with HF bridge. *Graph 1* HF path with a HF bridge, *Graphs 2 and 3* the first and the second sections of the HF path

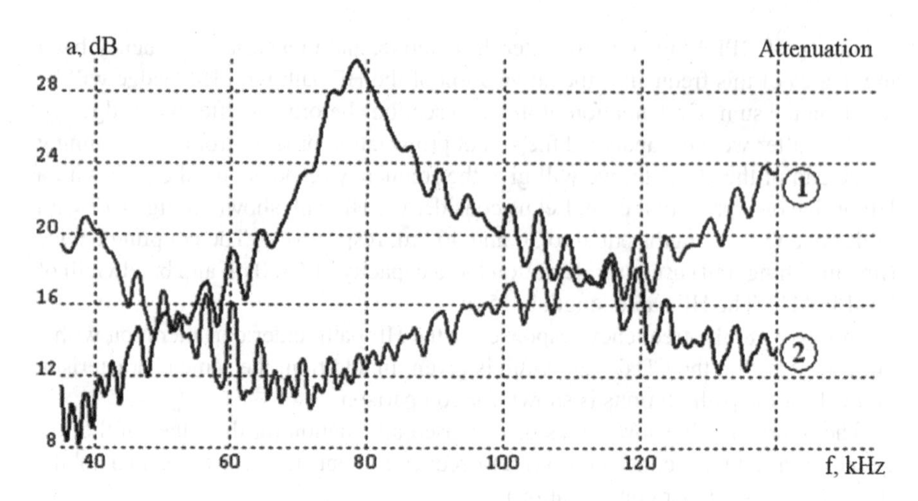

Fig. 4.19 Effect of changing the signal phase on one of the coupling devices in HF bridge scheme. *Graph 1* phase did not change (repetition of curve from Fig. 4.13), *Graph 2* phase changed

4.1.7 Features of Frequency Response of the HF Path Along Double-Circuit Overhead Lines

Hanging of double-circuit overhead line circuits can be carried out either on different towers (each circuit on "own" towers) or on double-circuit towers. Further, when mentioning double-circuit overhead lines, we will keep in mind the second case – hanging of both circuits on a common double-circuit tower, at least at most part of the line route.

Currently, such double-circuits 110 to 220 kV power lines are quite widespread in which circuit structures are different for different circuits. The most common overhead lines of this type are:

- Overhead line in which the intermediate substation is "cut in" into one of the circuits of a double-circuit line with outdoor switchgear located close to the overhead line route (Fig. 4.20a). In this case, it is possible that the intermediate substation does not have any PLC equipment or it has an HF bridge for PLC channels through the second circuit.
- OPL in which outdoor switchgear of the intermediate substation is located far from the main double-circuit overhead line path and connection to this substation is realized by two overhead lines hanged on separate towers or on double-circuit towers (Fig. 4.20b).
- OPL with a tap line connected to one of the circuits (Fig. 4.20c).

Parameters of HF paths created using such overhead lines have special features unlike the case when both circuits have the same structure over the entire length of the overhead line. These features are most noticeable in the paths along the circuit, in which there is no formal distortion of homogeneity (in Fig. 4.20 – OPL1).

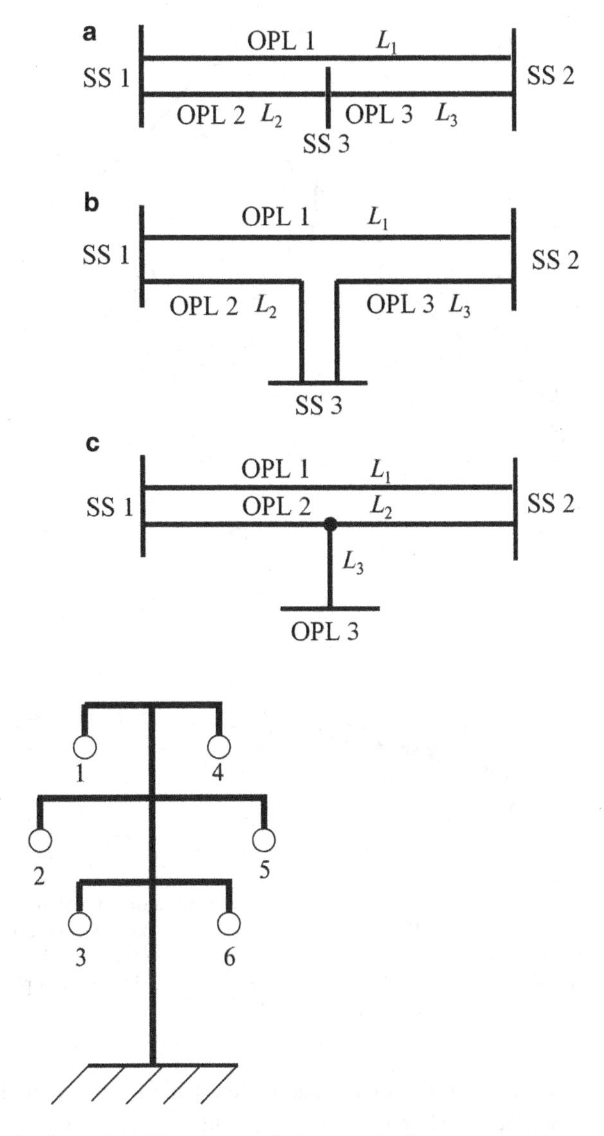

Fig. 4.20 Options of double-circuit overhead lines suspended on double-circuit towers with different circuit schemes

Fig. 4.21 Drawing of a double-circuit overhead line tower and phase numbering system used

However, this distortion is present since the line of the second circuit overhead line 2 with distorted homogeneity is part of an integrated six-wire overhead line.

In connection with the consideration of double-circuit overhead lines, let us say a few words about modal parameters of such overhead lines and optimal and non-optimal schemes for coupling to them.

Drawing of a double-circuit tower showing a phase numbering system used is given in Fig. 4.21.

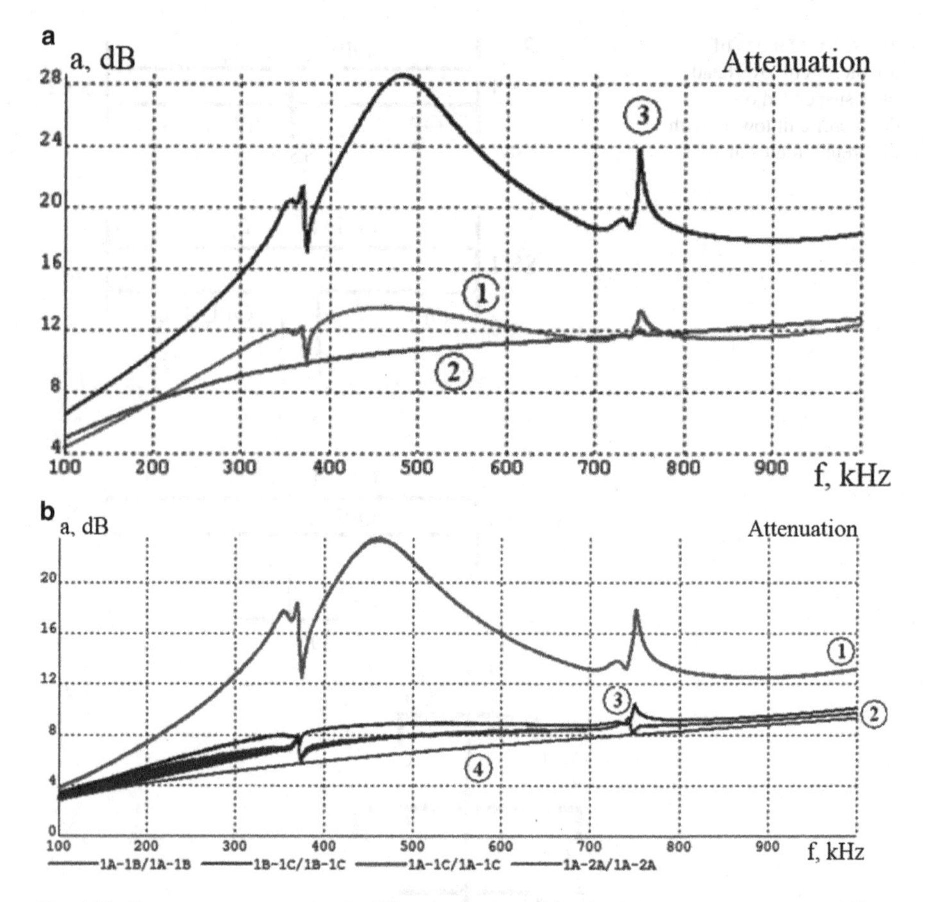

Fig. 4.22 Frequency response for the HF paths along double-circuit homogeneous overhead line: (**a**) – phase-to-earth, *Graph 1* phase 1 (or 4)-to-earth, *Graph 2* phase 2 (or 5)-to-earth, *Graph 3* phase 3 (or 6)-to-earth. (**b**) – phase-to-phase, *Graph 1* phase 1-to-phase 3 (or phase 4-to-phase 6), *Graph 2* phase 2-to-phase 2 (or phase 4-to-phase 5), *Graph 3* phase 2-to-phase 3 (or phase 5-to-phase 6), *Graph 4* phase 1-to-phase 4

From the set of modal parameters which was used in the description of the overhead line with horizontal phase arrangement (see Sect. 4.1.1), we will consider only the values of the matrix parameters λ, λ^{-1}, and γ. Given values of estimated parameters were calculated with the WinTrakt program taking into account periodic homogeneity distortions in the OPL installed on the towers on which OPL phase conductors are hanged. Not considering of such homogeneity distortions, in some cases, leads to a discrepancy between calculated values of the HF path attenuation and its actual values. Let us consider an example of modal parameters for double-circuit 220 kV overhead lines with a typical aluminum conductor steel reinforced (ACSR) cable of a phase conductor 19.5 mm in diameter.

Table 4.8 Attenuation constant α, phase constant β, and propagation velocity υ of modes of double-circuit 220 kV OPL

Frequency, kHz	Conditional name of mode					
	Mode 1 (−) Upper + center – lower		Mode 2 (+) Upper + lower – center		Mode 3 (−) Upper + center – lower	
	α	β/υ	α	β/υ	α	β/υ
50	0.027	1.0507/299	0.022	1.0508/299	0.028	1.0505/299.1
100	0.039	2.0999/299.2	0.039	2.0996/299.3	0.04	2.0991/299.3
200	0.055	4.1972/299.4	0.056	4.1966/299.4	0.057	4.197/299.5
420	0.083	8.8063/299.7	0.083	8.809/299.6	0.083	8.81/299.5
900	0.12	18.872/299.7	0.12	18.869/299.6	0.12	18.864/299.8
Mode 4 (+) Upper + center – lower		Mode 5 (−) Circuit-circuit		Mode 0 (+) Earth		
α	β/υ	α	β/υ	α		β/υ
0.037	1.05058/297	0.039	1.05072/296,9	0.67		1.1982/262.2
0.049	2.112/297.5	0.68	2.1126/297.4	1.14		2.3329/269.3
0.08	4.220/298	0.122	4.219/297.9	1.85		4.5635/275.4
0.14	8.854/298.1	0.22	8.8462/298.3	3.0		9.3943/280.9
0.25	18.96/298.2	0.4	18.932/298.7	4.85		19.859/284.8

Notes: units of matrix elements: α – dB/km; β – rad/km;
Unit of propagation velocity υ – thousand km/s

Further, the following is given:

- Matrix:

$$\lambda = \begin{pmatrix} 1 & 1 & 1 & 1 & 1 & 1 \\ 0,81 & -1,7 & -0,81 & 0,33 & 2,5 & 1,1 \\ -0,96 & 0,57 & 0,33 & -0,7 & 2,8 & 1,26 \\ -1 & 1 & -1 & 1 & -1 & 1 \\ -0,81 & -1,7 & 0,81 & 0,33 & -2,5 & 1,1 \\ -0,96 & 0,57 & 0,35 & -0,7 & 2,8 & 1,26 \end{pmatrix}$$

and its inverse matrix:

$$\lambda^{-1} = \begin{pmatrix} 0,188 & 0,149 & -0,2 & -0,188 & -0,149 & 0,2 \\ 0,118 & -0,2 & 0,078 & 0,188 & -0,2 & 0,078 \\ 0,282 & -0,227 & 0,104 & -0,282 & 0,227 & -0,104 \\ 0,281 & 0,071 & -0,284 & 0,281 & 0,071 & -0,284 \\ 0,03 & 0,08 & 0,097 & -0,03 & -0,08 & -0,097 \\ 0,1 & 0,128 & 0,207 & 0,1 & 0,128 & 0,207 \end{pmatrix}$$

and, to facilitate understanding, these matrices are not represented by their values obtained directly by calculation but in a somewhat simplified form (purely real and with rounded values of elements).

- Attenuation and phase constants, as well as the mode propagation velocities for several frequencies (Table 4.8).

While considering elements of λ and λ^{-1} matrices, a feature of these matrices' structure can be seen. This feature consists in the fact that the elements corresponding to phases of different circuits, symmetrically located on the tower, have the same signs for three modes (such modes are conventionally designated with + sign) and another three – different signs (such modes are conventionally designated with – sign). This can be interpreted in the sense that when a mode is excited in the line, the direction of the voltage vectors (currents) in the phase conductors of different circuits symmetrically located on the tower coincides for mode (+) and is opposite for mode (–).

As can be seen from the data given in Table 4.8, a six-mode double-circuit OPL, in accordance with modal theory, has one earth mode and five interphase modes:

1. Three interphase modes with approximately the same propagation constants. These are two (–) modes and one (+) mode:
   - Mode (–): upper + center phases – lower phase
   - Mode (–): lower phases – center phase
   - Mode (+): upper + lower phases – center phase

These modes' attenuation constant is approximately proportional to the square root of the frequency (is determined practically only by the losses in the phase conductors). The propagation velocity of the wave with these modes is more than 299 thousand km/s (close to light speed). This group of modes can be called modes with minimal losses.

2. Two interphase modes whose propagation constants, along with losses in the conductors, are affected by earth losses. Therefore, attenuation constants of these modes are more, and the propagation velocity is less than that of the modes with minimal losses considered above.

Due to different effects of earth mode, the propagation constants of the two modes are different.

One of these modes is a (+) mode and another – a (−) mode:

- Mode (+): upper + center phases – lower phase
- Mode (−): circuit to circuit mode

This group of modes can be called modes with increased losses.

As for the selection of optimal coupling schemes, for a non-transposed overhead line, in the case when the structure of both circuits throughout the overhead line is the same, they are "quasi-single-mode" (or optimal) coupling schemes:

- Phase 2 (center)-to-earth. Attenuation of the HF path which uses these schemes increases monotonously with frequency (without any signs of the attenuation pole). Signal propagation along the line built in accordance with this scheme occurs almost only with modes having minimal losses.
- Phase 1 (upper)-to-earth frequency response of the HF path built in accordance. This scheme has a weakly visible attenuation pole which is due to the fact that, in addition to the minimal loss mode, a (+) mode with high losses, having smaller propagation velocity than a mode with minimal losses, takes noticeable part in the signal propagation.
- Phase 1-to-phase 2, phase 2-to-phase 3, and phase 1-to-phase 4. Attenuation of the HF path which uses these schemes increases monotonously with frequency (without any signs of the attenuation pole). Signal propagation along the line built in accordance with this scheme occurs almost only with modes having minimal losses.

It should be noted that calling the scheme "phase 1 (upper)-to-earth" optimal would be stretching a point only for paths of double-circuit overhead lines, in which the structure of both circuits is the same. As shown below, if the circuit structures are different, this scheme cannot be called optimal.

To illustrate the said above, in Fig. 4.22, the frequency response for the HF paths along double-circuit non-transposed overhead lines with a length of 65 km are given for different coupling schemes.

Here, features of the HF path frequency response will be illustrated for a model in which ends of both circuits are loaded at approximately matching impedance and the signal source and receiver are connected directly to the OPL phase conductors.

This is done to exclude influence of reflected waves and coupling devices with line traps on frequency response of the HF path and to reveal to the greatest extent influence on processes in the OPL itself on the frequency characteristic.

Figure 4.22 clearly shows the difference in the frequency characteristics for optimal and non-optimal schemes of coupling to the overhead line, reasons for which were discussed in Sect. 4.1.2. In Fig. 4.22, local attenuation change, common for double-circuit overhead line paths, near the so-called resonant frequencies of the span is also clearly visible. These changes are most evident for suboptimal schemes of coupling to the overhead line. They are caused by influence of periodic distortion of the homogeneity of the overhead line near the towers on conditions for the signal propagation along the overhead line.

Resonant span frequencies $f_{res.span}$ correspond to frequencies for which the span length L_{span} is a multiple of half their wavelength. They are defined as:

$$f_{res.span} \approx \frac{300}{2L_{span}} \kappa$$

where $\kappa = 1,2..$, kHz

Used for calculation span length 0.4 km, these span resonant frequencies are equal to 375 and 750 kHz.

"Severity" of local attenuation increase near resonant frequencies is caused by the spread of span lengths over the overhead line length. The smaller this spread, the more "pronounced" attenuation increase at span resonant frequency and the narrower the frequency range in which the local attenuation increase occur.

Now, let us return to the parameters of paths along the double-circuit lines' structure of which is shown in Fig. 4.20.

Let us consider features (in comparison with the above case when both OPL circuits are the same all over the line) of the attenuation frequency dependence for the HF path along OPL1 of the first circuit for different structures of circuit 1 and circuit 2 of the double-circuit line.

Variant 1 OPL scheme shown in Fig.4.20a

Here we will analyze the path along the first circuit when there is the second circuit of the intermediate substation (SS3). SS3 is connected to the second circuit at 35 km from SS1. SS3 is represented in the analysis by the equivalent capacitor with a capacity equal to the sum of the capacities of high-voltage equipment installed on the outdoor switchgear of considered voltage. As we have said above, this capacity can be assumed to be equal to 5000 pF.

For the case in question, Fig. 4.23 shows the HF path frequency response for the paths with phase-to-earth coupling and, for comparison, calculation results for the case of second circuit without intermediate substation 3.

As can be seen from Fig. 4.23, the HF path frequency response change occurs almost only for non-optimal coupling schemes (which, as we have already noted, includes phase 1-to-earth scheme). This is, mainly, frequency shift of the

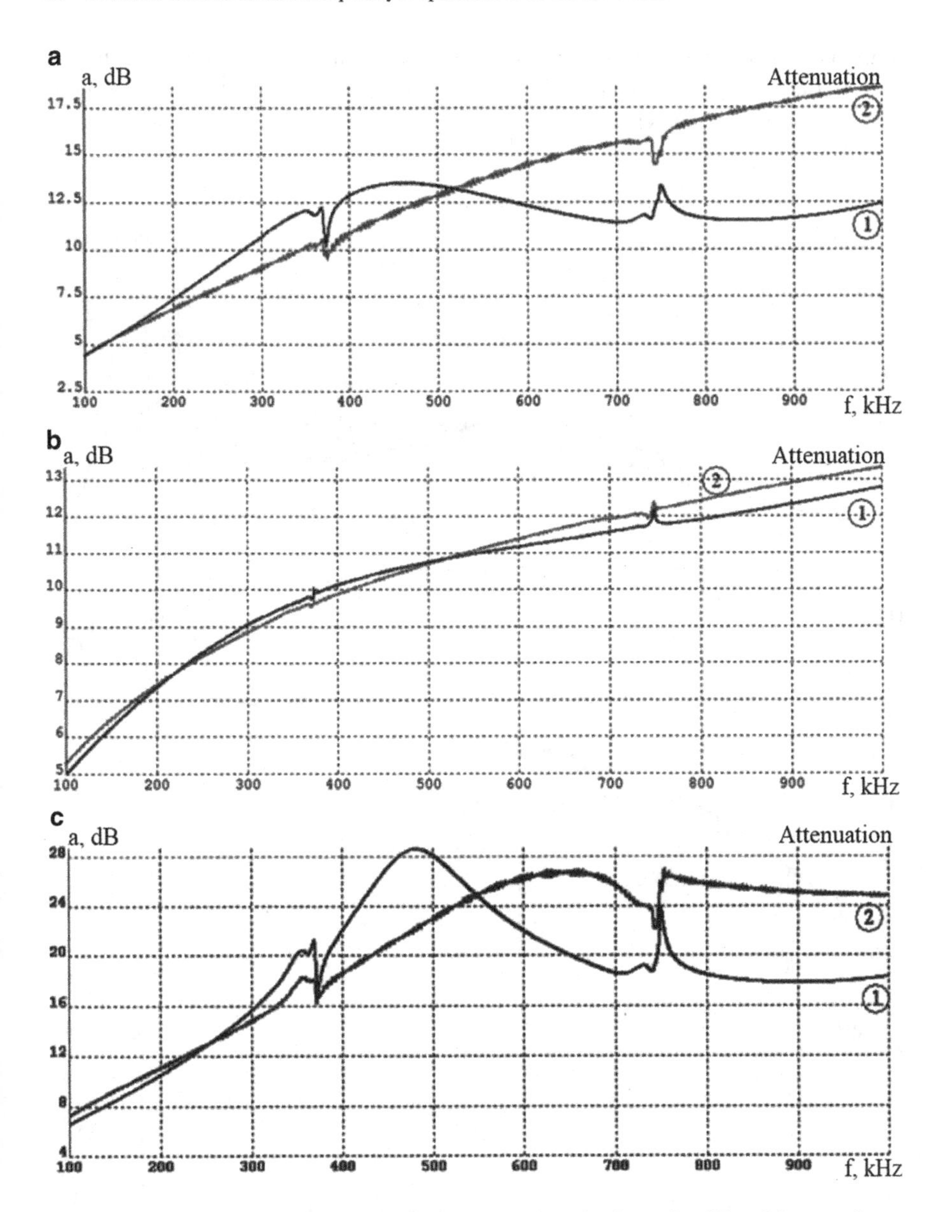

Fig. 4.23 Frequency response of the HF path along the first circuit overhead line: (**a**) upper phase-to-earth; (**b**) center phase-to-earth; and (**c**) lower phase-to-earth. *Graphs 1* without intermediate substation 3 in the second circuit, *Graphs 2* with intermediate substation 3 in the second circuit

attenuation pole into the area of higher frequencies. The frequency shift is due to changes in the phase angles between vectors of the incident wave voltage for the first and second circuit phases after the place of homogeneity distortion in the second circuit (SS3). This change, in turn, leads to corresponding change of the amplitude–phase ratios of mode voltage vectors and, as a result, to the attenuation pole frequency shift and change of the "severity" of the attenuation rise for non-optimal coupling schemes.

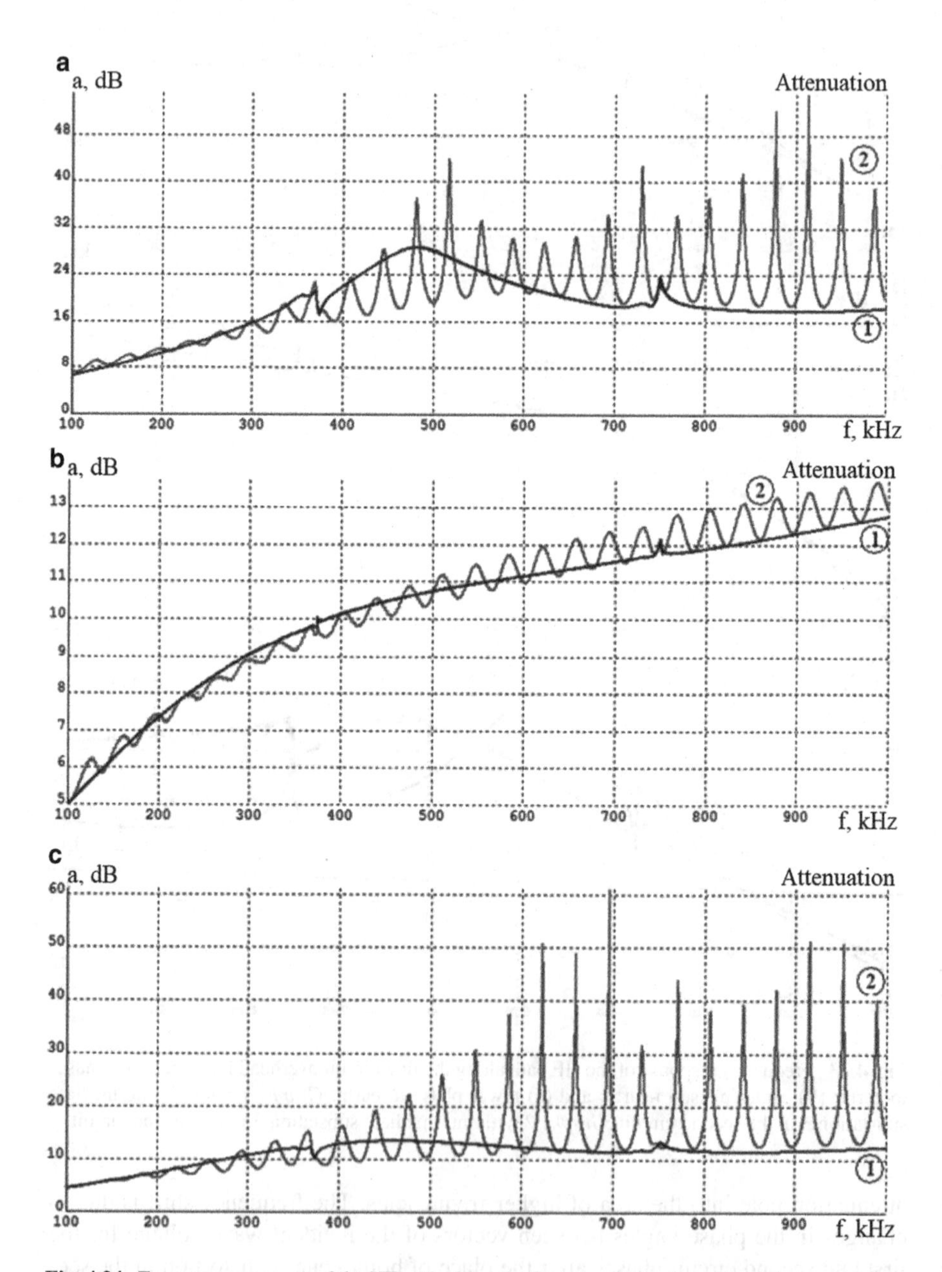

Fig. 4.24 Frequency response of the HF path along the first circuit overhead line: (**a**) upper phase-to-earth; (**b**) center phase-to-earth; and (**c**) lower phase-to-earth. *Graphs 1* without intermediate substation 3 in the second circuit, *Graphs 2* with intermediate substation 3 connected to the second circuit by the incoming line

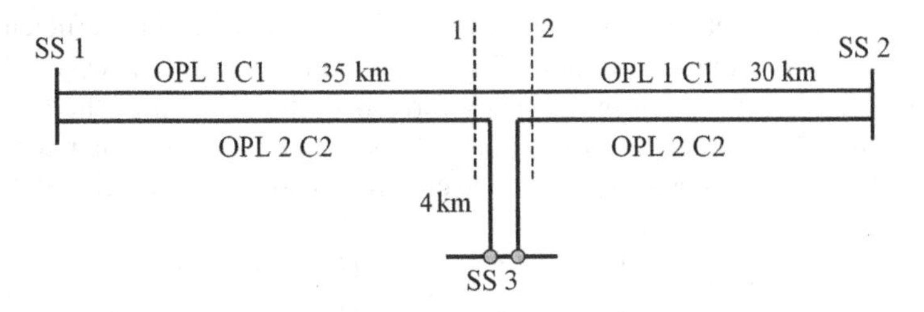

Fig. 4.25 A special case of the HF path scheme for a double-circuit overhead line

Thus, for non-optimal coupling schemes for connecting to the first circuit, adding the intermediate substation into the second circuit reduces attenuation of the path in one frequency range and increases it in another range.

Variant 2 OPL scheme shown in Fig.4.20b

Here, the previous option is considered, but for the case when the intermediate substation (SS3) is 4 km away from the main overhead line route, it is connected to the section of the second circuit using the incoming overhead line. For further analysis, the type of the overhead line hanger (double-circuit or separate support) does not matter in principle.

In Fig. 4.24, the frequency response of the paths for the case under consideration when the path is connected to the overhead line using different phase-to-earth coupling schemes is given. For comparison, this figure shows calculation results for the case with no intermediate substation 3 in the second circuit.

It can be seen from Fig. 4.24 that, similar to previous case, the HF path frequency response change occurs almost only when non-optimal coupling schemes are used. However, this change in nature is completely different. It consists in appearance of periodic change of attenuation between maximum and minimum values.

To understand what causes these types of frequency characteristics, we will consider the signal transmitting conditions over the overhead line between points 1 and 2 in the diagram Fig. 4.25 that is a special case of the diagram in Fig. 4.20b.

The signal transmission direction from substation 1 to substation 2 will be considered.

Phases voltage vectors in point 1 for both circuits, provided that there are no waves reflected from point 1, are defined by a linear combination of the voltages of the incident waves of interphase modes involved in signal propagation which is defined by the formula (4.4) written in matrix form, which in this case is:

$$U_{\text{ph1}} = \lambda U_{\text{m1}}$$

where U_{ph1} is a complex column matrix of the 6th order containing phase voltage vectors in point 1 and U_{m1} is a complex column matrix of the 6th order containing incident wave voltage vectors in modes in point 1 which are related with the excited at the beginning of the double-circuit line on SS1 by the formula (4.6).

Voltage vectors of phases of both circuits in point 2 are also represented by a complex column matrix of the 6th order U_{ph2}.

The relationship between the matrices U_{ph2} and U_{ph1} at a known coefficient T of voltage transfer from point 1 to point 2 is written in matrix form as $U_{\text{ph2}} = T\,U_{\text{ph1}}$. This formula can be written in the block form with the division of the 6th order matrix of the OPL voltage vectors in points 1 and 2 into two block matrices of the 3rd order containing voltage vectors of the circuit 1 phases (U_{c1}) and circuit 2 (U_{c2}):

$$\begin{pmatrix} U_{c12} \\ U_{c22} \end{pmatrix} = \begin{pmatrix} ''_{11} & ''_{12} \\ ''_{21} & ''_{22} \end{pmatrix} \begin{pmatrix} U_{c11} \\ U_{c21} \end{pmatrix} \tag{4.10}$$

To describe the T_{11}, T_{12}, T_{21}, and T_{22} blocks of the voltage transfer coefficient matrix T, we will make some assumptions which significantly simplify the issue, but do not affect its essence:

- For circuit 1, we will assume that the phase voltage vectors in points 1 and 2 are pairwise equal.
- Lines in the section of line incoming from the main line to substation 3 are perfect (without losses).
- The SS3 capacity does not cause an effect on conditions of wave propagation along the incoming lines, and the same-named phases of the incoming line at the substation 3 are simply connected in pairs.

Taking these assumptions into account, the formula (4.10) which defines relationship between the voltage vectors of the first and second circuit phases in point 2 and the voltage vectors of the same circuit phases in point 1 can be written as:

$$\begin{pmatrix} U_{c12} \\ U_{c22} \end{pmatrix} = \begin{pmatrix} 1 & 0 \\ 0 & \exp(-j\phi) \end{pmatrix} \begin{pmatrix} U_{c11} \\ U_{c21} \end{pmatrix}$$

where U_{c11} and U_{c12}, complex column matrix of the 3rd order containing phase voltage vectors of the 1st circuit in the points 1 and 2, respectively; U_{c21} U_{c22}, complex column matrix of the 3rd order containing phase voltage vectors of the 2nd circuit in the points 1 and 2, respectively; 1, diagonal identity matrix of 3rd order, representing block T_{11} of matrix T; 0, zero-diagonal matrix of the 3rd order representing blocks T_{12} and T_{21} of matrix T; and $\exp(-j\varphi)$, diagonal matrix of the 3rd order representing block T_{22} of matrix T and defining shift angle of the second chain phase voltage on doubled length of the incoming line $L_{\text{input.line}}$:

$$\varphi \approx \frac{4\pi f L_{\text{input.line}}}{300}$$

Taking into account the formula (4.1), the angle φ is calculated as rad.

When the frequency changes, the angle U_{ph1} periodically passes all values from 0 to 2π rad (or from 0 to 360°).

It can be obtained from (4.10) that the frequency interval corresponding to angle φ change by 2π (180°) is defined by the formula:

$$\Delta f \approx \frac{300}{2L_{\text{input.line}}} \tag{4.11}$$

For a frequency at which $\varphi = 0°$ or $360°$ ($\exp(0) = \exp(-2\pi) = 1$), the formula (4.10) takes the form:

$$\begin{pmatrix} U_{c12} \\ U_{c22} \end{pmatrix} = \begin{pmatrix} 1 & 0 \\ 0 & 1 \end{pmatrix} \begin{pmatrix} U_{c11} \\ U_{c21} \end{pmatrix} \text{or} \left|U_2\right| = \left|U_1\right| \tag{4.12}$$

that is, the sixth-order matrices containing voltage vectors of the double-circuit overhead line phases in points 1 (U_1) and 2 (U_2) will be the same, and modal components of the wave propagating further from point 2 toward substation 2 coincide with the incident wave in point 1 coming from substation 1, and, consequently, the HF path attenuation between substation 1 and 2 will be approximately the same as in the absence of a line incoming to substation 3.

For a frequency at which $\varphi = 180°$ ($\exp(-\pi) = -1$), the formula (4.12) takes the form:

$$\begin{pmatrix} U_{c12} \\ U_{c22} \end{pmatrix} = \begin{pmatrix} 1 & 0 \\ 0 & -1 \end{pmatrix} \begin{pmatrix} U_{c11} \\ U_{c21} \end{pmatrix}$$

that is, block matrices of the third order containing voltage vectors of the circuit 2 phases in points 1 (U_{c21}) and 2 (U_{c22}) will have different signs. This can lead to excitation in point 2 of modal components of the wave propagating further from point 2 toward substation 2 with such amplitude and phase relations which, under certain conditions (see Sect. 4.1.2), lead to appearance of attenuation poles in the frequency response of the HF path.

Thus, in accordance with (4.10), the period of the attenuation pole occurrence in this case (at $L_{\text{input.line}} = 4$ km) is approximately 37 kHz that is confirmed by the data given in Fig. 4.24.

Variant 3 OPL scheme shown in Fig.4.20c

Here we will analyze the path along the first circuit when there is the second circuit of the branch substation (SS3). The frequency response of the HF paths with coupling according to phase-to-earth schemes for this case is shown in Fig. 4.26. For comparison, this figure shows calculation results for the case with no tap line in the second circuit. It can be seen from Fig.4.26 that the frequency response change occurs almost only when non-optimal coupling schemes are used. This is a rather large influence of the tap line located in the second circuit on attenuation of the HF path along the first circuit.

A simplified consideration of the frequency response in this case can be carried out by analogy with the previous case. At that, the tap line can be replaced by a complex impedance *Z*(*f*) located in the appropriate point between each phase of the

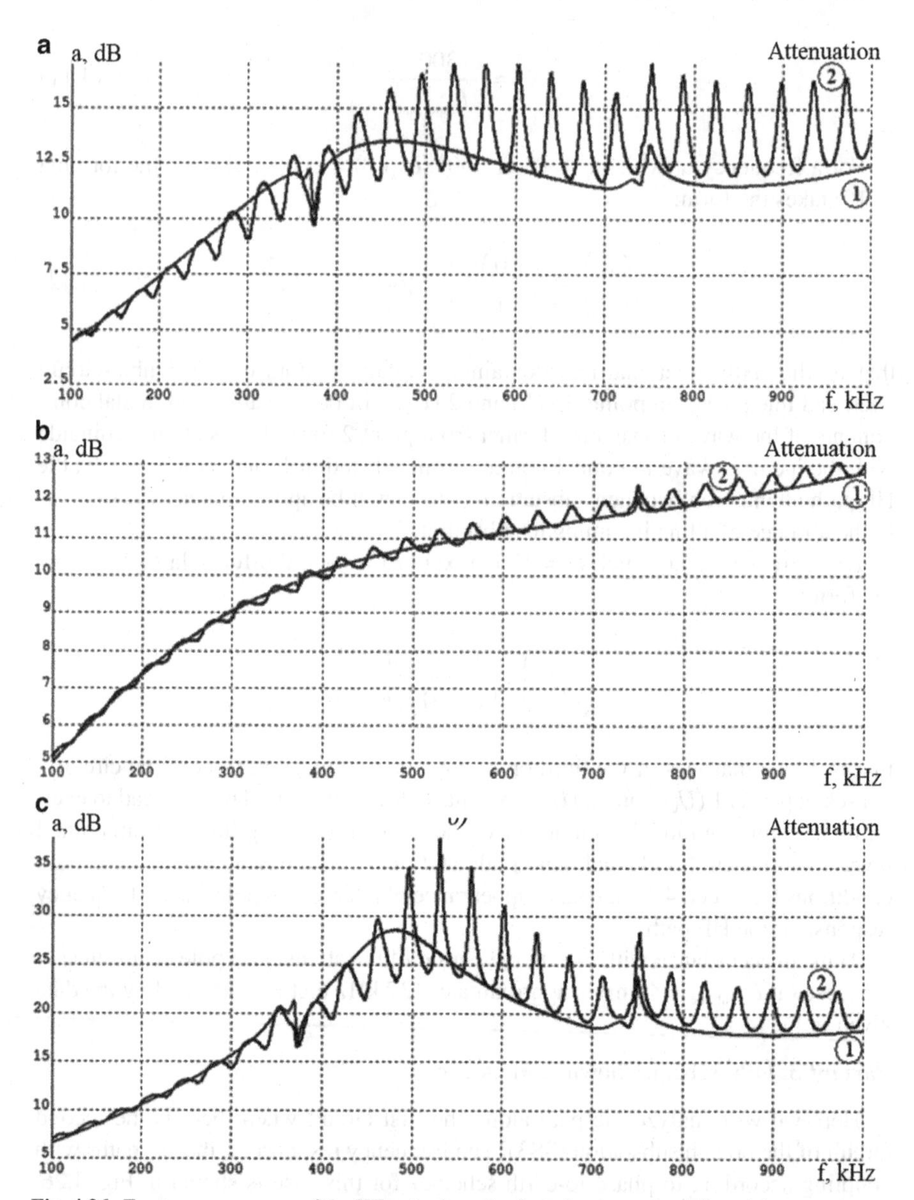

Fig. 4.26 Frequency response of the HF path along the first circuit overhead line: (**a**) upper phase-to-earth; (**b**) center phase-to-earth; and (**c**) lower phase-to-earth. *Graph 1* without intermediate substation 3 in the second circuit, *Graph 2* with line to intermediate substation in the second circuit

second circuit and earth. The value of this impedance modulus and its real and imaginary parts depends on frequency, and this dependence is periodic. The period is defined by the formula (4.11) in which $L_{\text{tap line}}$ (length of the tap line) is used instead of $L_{\text{input.line}}$.

4.1.8 *Frequency Response of the HF Path Along Cable and Cable-Overhead Power Lines*

Cable and mixed cable-overhead power transmission lines (respectively, CPL and COPL) with the use of low- and high-pressure oil-filled cables or with insulation made of cross-linked polyethylene (XLPE) are quite widely used in 35 to 220 kV power grids.

In continuation of the previous sections in which features of the influence of various factors on parameters of the HF path along overhead power line (non-transposed or transposed), we will consider here features of frequency response of the HF paths along CPL and COPL (performing calculations using the WinTrakt software). The theoretical basis for calculating the CPL and COPL wave parameters is presented in [7, 13].

Since this feature analysis will be carried out using the modal theory, we will first describe modal parameters of a three-phase CPL each phase of which is a coaxial cable with cross-linked polyethylene insulation.

According to provisions of the modal theory, such a CPL (three phases and three sheaths) is characterized by six modes. The analysis shows that these modes can be divided into three groups by values of their propagation constants.

The first group includes three modes which have the same propagation constants, i.e., attenuation constant and phase constant (and propagation velocity related to it by the formula (4.1)). The attenuation constant is defined only by losses in the conductors' resistance and sheath losses in the inner cable insulation (i.e., dielectric loss tangent tg δ), it is approximately proportional to the square root of frequency, and propagation velocity which defines the phase constant is close to 200 thousand km/s (related with light speed by coefficient $\sqrt{1/\varepsilon}$, where ε is relative permittivity of the inner cable insulation.

These mode parameters do not depend on the cable relative arrangement in earth. Modes included in this group can be called coaxial (intra-cable).

The second group includes two modes which have similar (but still different) propagation constants which are lower than those of intra-cable modes. The

Table 4.9 Attenuation constant α, phase constant β, and propagation velocity υ of 110 kV CPL modes with a trefoil cable arrangement in earth

Frequency, kHz	Conditional mode name							
	Intra-cable Modes 1, 2, and 3		Sheath mode 4, No.1		Sheath mode 5, No.2		Earth-mode Mode 0	
	α	β/υ	α	β/υ	α	β/υ	α	β/υ
50	0.163	1.6737/187.7	0.079	3.6894/85.1	0.069	4.267/73.6	8.4	13.055/24.1
100	0.23	3.3369/188.3	0.13	7.3745/85.2	0.11	8.5307/73.7	18.3	25.26/24.9
200	0.34	6.6589/188.7	0.21	14.743/85.2	0.175	17.057/73.7	38.4	48.724/25.7
400	0.5	13.297/189	0.37	29.475/85.3	0.29	34.106/73.7	84.9	93.604/26.8
800	0.74	26.564/189.2	0.723	58.932/85.3	0.49	68.203/73.7	188.4	178.87/28.4

attenuation constant is approximately proportional to frequency, and the propagation velocity is within 50 to 80 thousand km/s.

These mode parameters depend on the cable relative arrangement in earth. Modes included in this group can be called sheath modes (sheath mode 1 and sheath mode 2).

The third group includes the remaining sixth mode. It has the highest attenuation constant and the lowest propagation velocity of approximately 25 to 30 thousand km/s. Its parameters also depend on relative cable arrangement in earth. This mode is called earth or zero mode.

To have an idea of modal attenuation constants (α) and propagation velocities (υ), values of these parameters for 110 kV CPL with a triangular phase arrangement in earth with a depth of about 1 m and distances between the phase cable axes of about 0.2 m are given in Table 4.9.

Wave impedance of intra-cable mode which will be necessary in the future is equal to the wave impedance of the coaxial cable calculated by the formula:

$$Z_{\text{input.CPL.}} = \frac{138}{\sqrt{\varepsilon_r}} \lg\left(\frac{D_{\text{sheath}}}{D_{\text{core}}}\right) \tag{4.13}$$

where D_{sheath} is inner diameter of the sheath; D_{core} is diameter of the core conductor; and ε_r is relative permittivity of the inner insulation (usually is within 2.3 to 2.5). For the cable type considered above, the wave impedance of intra-cable mode is 27 ohms.

In further examples of frequency response for HF paths along CPL and COPL with different schemes in the cable line junction in the HF path scheme, this type of cable will be used (to obtain comparable results for different calculations).

Before proceeding to the analysis of features of frequency response for the HF path along CPL and COPL, let us consider features of the implementation of the coupling device with coupling capacitor and line trap used to create PLC channels along cable lines. They are as follows:

- Standard line traps, reactors, and coupling capacitors are used, since their parameters do not depend on the type of power line (CPL or OPL) used.
- Coupling device and tuning elements of line traps (which may have a standard schema, the same for OPL and CPL) should conform to the standard which specify the numerical value related to characteristic impedance of CPL (return loss of the coupling device at the line and HF cable sides should be not less than 12 dB); blocking impedance of LT should be not less than 1.41 times more than the CPL characteristic impedance.

The calculated OPL characteristic impedance value for 35 to 220 kV OPL is equal to 450 ohms, and the characteristic impedance of the cable line is approximately equal to the wave impedance of the intra-cable mode (4.13), and the CPL with a voltage of 35 to 220 kV is 20 to 45 Om (average). Therefore:

- Line traps with standard tuning elements intended for overhead lines can be directly used for cable lines, since their blocking impedance will always be higher than the required for overhead lines.
- The coupling device unit for CPL should be especially designed. At that, the coupling device for CPL which has the CC capacity equal to that of the coupling device for OPL has a significantly narrower bandwidth.

Let us proceed to features of frequency response of HF paths along CPL and COPL.

Similar to the previous section, the illustration of features of the line path frequency response will be done for a model in which CPL or OPL is loaded at the end of the path to about equal matching impedance, and the signal source and receiver are connected directly to the phase connectors of CPL or COPL, without using coupling devices and line traps. We will repeat that this model significantly simplifies analysis of processes occurring in CPL and COPL themselves, without compromising their understanding.

Let us proceed to the features of attenuation characteristics for the HF path along CPL.

Design of the cable itself, soil resistivity, and arrangement of cables used for the CPL in earth (coordinates of the cable axes), of course, affect conditions for HF signal propagation over the CPL. However, the most important effect on the frequency response of the HF path along CPL and COPL is caused by the method of the cable sheath earthing (at one or both ends of the CPL) and the presence of transpositions of the cable sheaths. The effect of these factors will be considered further.

First, let us consider the features of frequency response without sheaths' transposition for the HF path along CPL, provided that the sheaths are earthed at both ends of the CPL, and then – the change of these characteristics when the sheaths are earthed only at one CPL end or sheaths' transposition is performed.

Variant 1 CPL without sheath transposition

CPL sheaths are not transposed only if CPL length is small (about 200 to 300 m). Because of this, in practice, it is unlikely to meet PLC channel over a CPL with such a small distance between substations. Nevertheless, this analysis is interesting methodically for understanding the effect of the sheath earthing scheme on attenuation of the line path along CPL, as well as when considering cable junction into the overhead line which may have a small length.

If there is no sheath transposition and the sheaths are earthed at both CPL sides, only coaxial (intra-cable) modes participate in signal propagation.

In this case, such a combination of intra-cable modes is excited that the signal propagates only over the cable to which the coupling is made. Therefore, soil resistivity and relative arrangement of cables of different phases in the earth do not affect parameters of the line path.

Figure 4.27 shows the frequency response of the line path arranged according to the phase-to-earth coupling scheme along CPL with the length of 200 m and with trefoil CPL phase arrangement in earth.

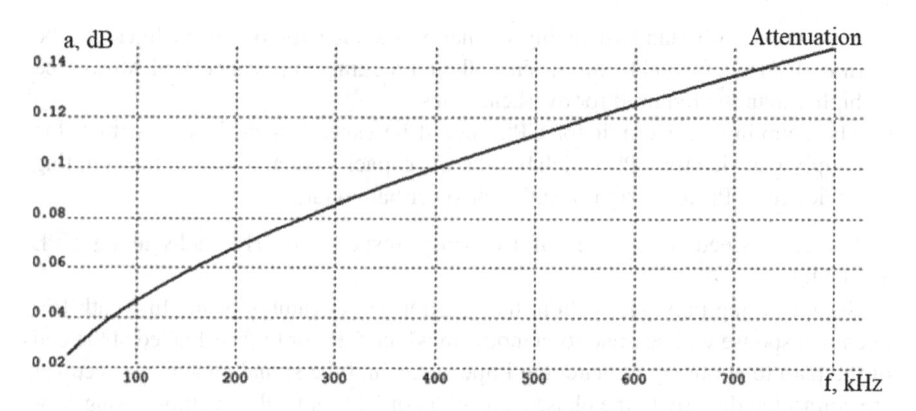

Fig. 4.27 Frequency response of the line path along 200 m long CPL when the sheaths are earthed at both sides

As can be seen from Fig. 4.27, the path attenuation changes smoothly with a frequency that is typical for single-mode coupling schemes. The attenuation value is equal to the attenuation constant of the intra-cable mode (see Table 4.9) multiplied by the CPL length (0.2 km). If the sheaths are isolated at one CPL end (insulation is used to reduce losses and heating in the sheaths), the line path parameter behavior changes dramatically.

Figure 4.28 shows the frequency response of the line path which uses a scheme for which the frequency response is shown in Fig. 4.27 but with sheaths isolated at the left side of the CPL.

As can be seen from the comparison of Figs. 4.27 and 4.28, the sheath isolation at one end of the CPL led to a significant change in behavior of the frequency response of the line path. Conditions appeared for multimode transmitting signal over the CPL by all CPL modes, and non-homogeneities appeared in the path which led to reflected waves.

Variant 2 CPL with sheath transposition

For large CPL length, sheath transposition is applied that is intended (together with sheath insulation at one end of the CPL) for reducing currents in the sheaths and, accordingly, sheath heating.

One of possible sheath transposition schemes is shown in Fig. 4.29.

Let us consider the frequency response for the line path along 110 kV CPL, with the sheath transposition scheme shown in Fig. 4.29. The CPL length is 6 km (i.e., the length of the CPL sections between the sheaths' transposition points is 2 km). Location of each phase of CPL in earth is at the corners of triangle.

The frequency response of such a line path for phase B (center)-to-earth coupling scheme is shown in Fig. 4.30. It should be noted that attenuation of the line path coupled according to phase A-to-earth and phase C-to-earth schemes does not differ from the one shown in Fig. 4.30. For comparison, the same characteristics for the case of no sheath's transposition are also shown in this figure.

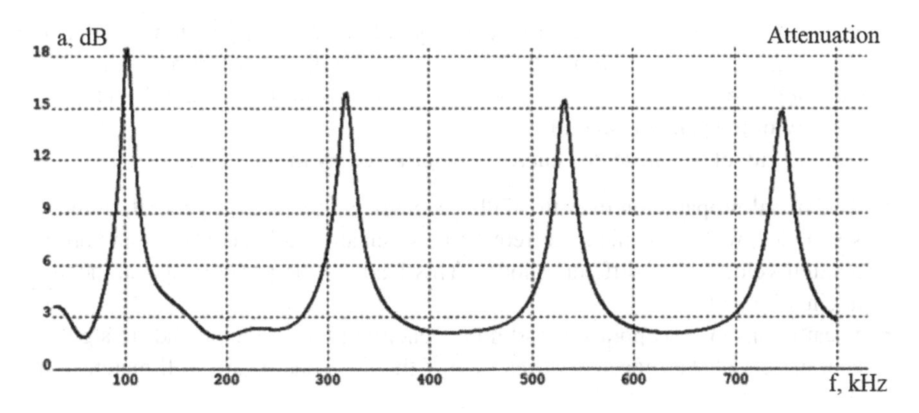

Fig. 4.28 Frequency response of the line path along 200 m long CPL when the sheaths are isolated at the left side and earthed at the right side

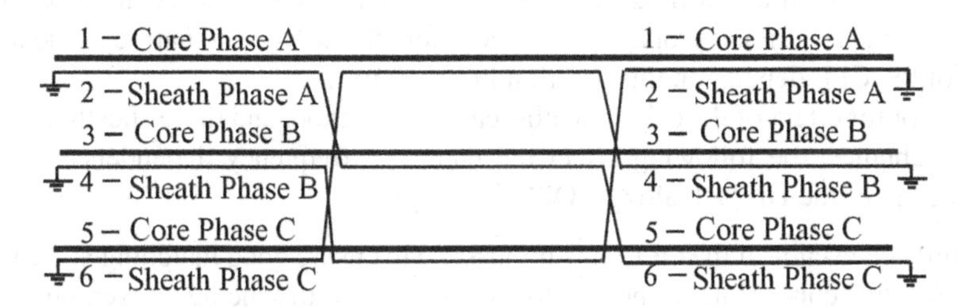

Fig. 4.29 Example of a diagram for CPL sheath transposition

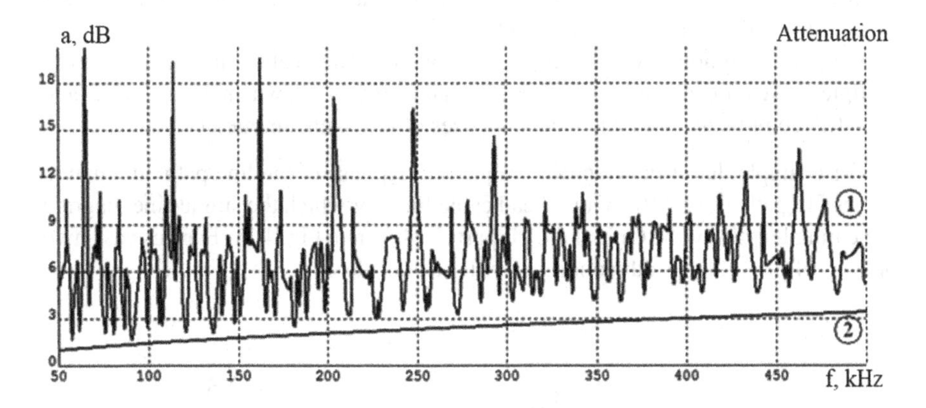

Fig. 4.30 Line path frequency response for CPL: *Graph 1* phase B-to-earth for CPL with sheath transposition, *Graph 2* phase B-to-earth for CPL without sheath transposition

It can be seen from the figure that the use of sheath transposition, similar to sheath isolation at one end of the CPL, leads to a significant distortion of the CPL homogeneity in the transposition points that are manifested in significant non-uniformity of the path attenuation.

Complexity of the presented curves is determined by the fact that:

- The signal propagation over the CPL becomes multimode (in general, with all six modes) with a significant difference between attenuation constants and propagation velocities of different modes. This leads to the periodic appearance of attenuation poles.
- Sheath transposition points distort homogeneity of the CPL that leads to significant effects of the reflected waves even if the CPL phases are well matched at its ends.

Variant 3 Mixed cable and overhead lines

The effect of the junction of the cable line in the overhead line depends on the nature of the process of signal propagation through the cable line (CPL). As shown above, the method of earthing the cable sheaths and sheath transposition schemes mostly influences signal propagation. In addition, the degree and nature of the influence of the CPL depend on this location in the COPL.

Except for effect of the cable sheaths, earthing method, and these sheath transposition schemes, the following affect the nature of frequency dependences of the parameters of the HF path along COPL:

- Unmatched ends of the OPL. It is impossible to ensure good matching at the OPL ends when connecting to the line using phase-to-earth scheme or even phase-to-phase. In this case, the path end reflection coefficient depends on the overhead line switching state, and disconnected and not earthed COPL can reach 0.7.
- Significant mismatch between the overhead line and the cable line at the junction point. The modulus of the reflection coefficient for interphase wave in these points equals to approximately 0.84.
- Lengths of homogeneous sections of the overhead and cable lines in which multiple waves are reflected are different, and this, as shown in Sects. 4.1.2 and 4.1.5, complicates frequency dependences of line path parameters.

For example, let us consider the frequency response of the line path attenuation for the HF path along 110 kV overhead cable line, for which the single-line diagram indicating lengths of the OPL and CPL parts is shown in Fig. 4.31. For this case, two options of the cable sheath conditions will be considered:

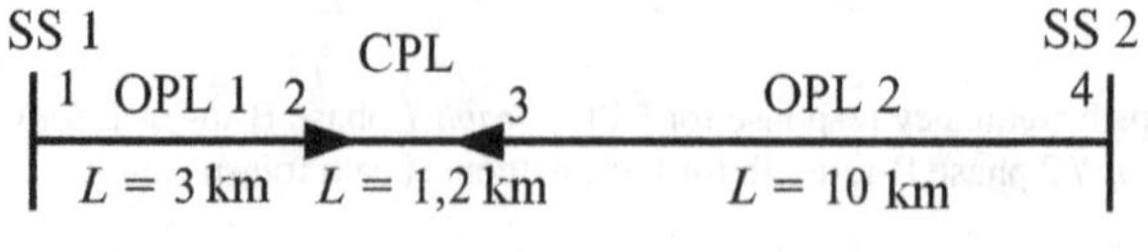

SS 1
CPL
SS 2
1 OPL 1 2
3
OPL 2
4
L = 3 km L = 1,2 km
L = 10 km

Fig. 4.31 Single-line COPL scheme for the example under consideration

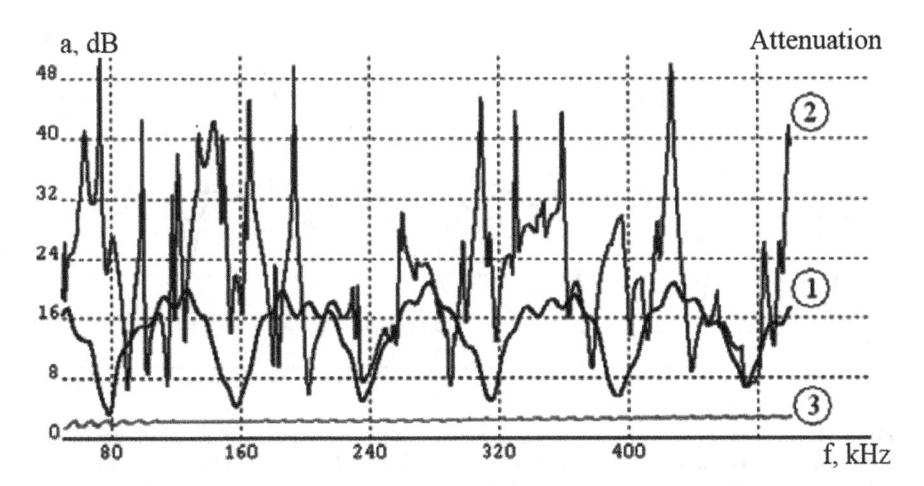

Fig. 4.32 Frequency response of the line path along COPL. *Graph 1* CPL sheaths without transpositions, *Graph 2* CPL sheaths are transposed, and *Graph 3* no cable junction

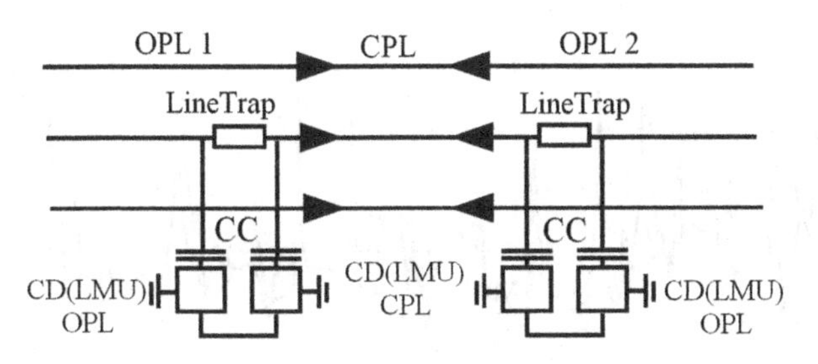

Fig. 4.33 Matching scheme for CPL and OPL connection

- When the sheaths are earthed at both ends of the CPL, without transposition of the sheaths
- When the sheaths are earthed at both ends of the CPL and are transposed according to the diagram given in Fig. 4.29

Figure 4.32 shows the frequency response of the line path along the COPL with a scheme shown in Fig. 4.31, for the two options of cable line sheath conditions mentioned above. In addition, the figure shows the frequency response for this path without a cable junction.

As can be seen in Fig. 4.32, an addition of a cable junction into the overhead line significantly changes the behavior of the HF path frequency response relative to the variant without the CPL. At that, as expected based on the above arguments, the "complexity" of the frequency response is most affected by the application of sheath transposition. The sheath isolation at one side of the cable line leads to approximately the same frequency response "complication" as in the case of sheath transposition.

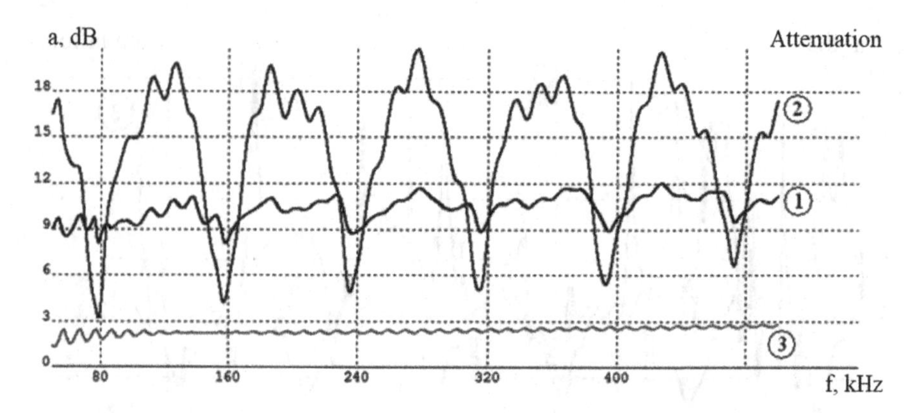

Fig. 4.34 Frequency response for the line path along COPL for the case of no sheath transpositions, provided that the sheaths are earthed at both ends. *Graph 1* with matching at the CPL and OPL connection, *Graph 2* without matching, *Graph 3* no cable line

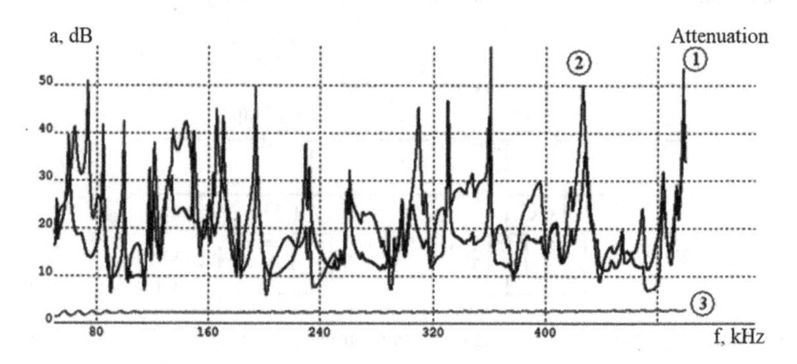

Fig. 4.35 Frequency response for the line path along COPL for the case of sheath transpositions and sheath earthing at both ends. *Graph 1* with matching at the CPL and OPL connection, *Graph 2* without matching, and *Graph 3* no cable line

It is recommended in [12] to include for CPL and OPL connection matching circuits according to the scheme similar to the HF bridge scheme, as shown in Fig. 4.33.

It should be noted that use of a matching scheme (see Fig. 4.33) improves the HF path parameters only if there is no CPL sheath transposition and sheath earthing at both ends of the cable. Figure 4.34 shows the frequency response for the line path which uses matching in the CPL and OPL connection points for the case of no sheath transposition and sheath earthing at both ends of the CPL. This figure also shows the frequency response for the same case but without matching devices. It can be seen in this figure how much the frequency characteristics improve when a matching is used in this case.

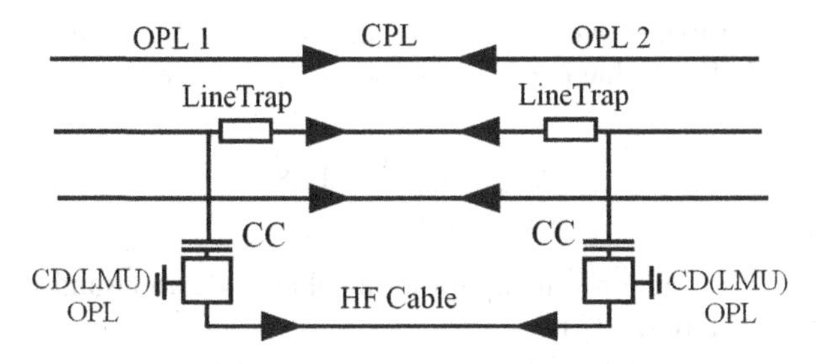

Fig. 4.36 HF bridging the entire cable line in the overhead line

Figure 4.35 shows the frequency response of the line path which uses matching schemes in the CPL and OPL connection points for the case of sheath transposition and sheath earthing at both ends of the cable.

This figure also shows the line path frequency response for the same case but without matching devices.

It can be seen in this figure that the use of matching circuits in the case of sheath transposition, although it changes the behavior of the frequency response of the line path, does not lead to its improvement.

In this case, a solution which uses the HF bridge of the entire cable line as shown in Fig. 4.36 can be used. At that, the frequency response of the line path with the HF bridge is quite satisfactory (close to that shown in Fig. 4.34).

However, when implementing the scheme given in Fig. 4.36, a problem of "electric potential removal" and related problems of reliability of the HF cable included in the HF bridge circuit and also safety of maintenance of the HF bridge circuit devices in the case of short currents in the electric grid should be considered. Ways to solve these problems are not considered here as they go beyond the topic under consideration.

The degree of the cable junction influence on the line path parameters, of course, depends on the cable length. The less ratio of cable length (L_{CPL}) to wavelength of the signal transmitted along the path (λ_{HFpath}), the less significant wave processes in the cable which lead to such a complex shape of frequency response of the HF path. If the (L_{CPL}/λ_{HFpath}) ratio does not exceed a certain value, the cable line with distributed parameters can be replaced in the design scheme with a capacitor with a capacity equal to that of the replaced cable.

To define the condition for such replacement possibility, the following formula can be used:

$$L_{\mathrm{CPL}} f_{\mathrm{max}} \leq K_{\mathrm{CPL}} \tag{4.14}$$

where L_{CPL} is the cable line length, km; f_{max} is the highest frequency of signals transmitted along the path, kHz; and K_{CPL} is a constant depending on permissible error in calculations and obtained when replacing the CPL with an equivalent capacity.

It can be shown that, if we assume a relative error in calculating the cable line impedance of 5%, K_{CPL} value in (4.14) can be assumed to be equal to:

$$K_{CPL} = \frac{0{,}37\upsilon}{2\pi} = 11.8$$

where υ is wave propagation velocity for intra-cable mode, thousand km/s, which is assumed to be 200 thousand km/s.

Capacity per unit length of the cable for the considered replacement can be defined according to the cable data sheet or by an approximate formula:

$$! = \frac{10^{-9}\varepsilon_r}{41.4} \lg\left(\frac{D_{sheath}}{D_{core}}\right)$$

where $D_{sheaths}$, D_{core}, and ε_r are the same values as in (4.13). This capacity is usually about 190 pF/m.

4.2 Return Loss and Input Impedance of the HF Paths

To ensure matching of the PLC equipment transmitter and the HF path, the PLC equipment transmitter impedance and input impedance of the HF path should be normalized. These impedance nominal values are assumed to be the same (75 ohms for phase-to-earth coupling scheme and 150 ohm for phase-to-phase coupling).

The modulus and phase angle of both input impedance of the HF path and internal impedance of the transmitter differ from nominal value, and this causes their mismatch. Mismatching of the transmitter internal impedance and input impedance of the HF path changes the operation mode of the transmitter that causes change of output power, increasing transmitter nonlinearity, and change in the transmitter thermal conditions.

The effect of these mismatch consequences depends on the mismatch degree and should be controlled.

The difference between the transmitter internal impedance value (Z_{TX}) and its nominal value (R_{nom}) is usually characterized by the transmitter's return loss ($a_{return.loss.TX.}$), which is defined as:

$$a_{return.loss.TX} = 20\lg\left|\frac{Z_{TX} + R_{nom}}{Z_{TX} - R_{nom}}\right| \tag{4.15}$$

The difference between the HF path input impedance value ($Z_{input\ .path}$) and its nominal value (R_{nom}) is usually characterized by the HF path return loss ($a_{return.loss.path}$), which is defined as:

$$a_{\text{return.loss.path.}} = 20\lg\left|\frac{Z_{\text{input.path}} + R_{\text{nom}}}{Z_{\text{input.path}} - R_{\text{nom}}}\right| \tag{4.16}$$

In (4.15) and (4.16), impedances Z_{TX} and $Z_{\text{input.path}}$ are represented by their full values (in complex form), and return loss is defined taking into account the difference in the impedances to be compared considering not only modulus but also their phase angle.

The more return loss, the closer internal impedance of the transmitter or input impedance of the HF path to their nominal values.

Frequency dependence of return loss of the HF path has a rather complex shape, defined by:

- Parameters of line traps and coupling devices for processing and connecting coupling and non-coupling phases
- OPL end conditions, i.e., loading of coupling and non-coupling phases at the ends of the line considering specified above in Sect. 4.1.2 (see Table 4.5) possible operation modes of the line
- Degree of influence of waves reflected from the homogeneity distortion points of the OPL (coefficient of wave reflection form homogeneity distortion points and attenuation of interphase waves along the length of the line section in which reflections occur)

Considering features of frequency dependence of return loss (and module of the HF path input impedance) will be carried out with the example of a simple HF path (including just one OPL) for two edge cases of these parameter frequency response behaviors.

The first edge case corresponds to a HF path along a long line with a sufficiently large attenuation of the HF path and, as a result, negligible influence of waves repeatedly reflected from the ends of the line.

Frequency dependence $a_{\text{return.loss.path}}$ and input impedance of this HF path (within nominal frequency band of the equipment) are "quasi-smooth." It is obvious that in this case, the transmitter matching (if it is required) should be carried out for the input impedance of the HF path corresponding to its average value within the nominal transmission frequency band.

The second edge case corresponds to the HF path along a short line with significant influence of reflected waves. Frequency dependence $a_{\text{return.loss.path}}$ and input impedance of the HF path in this case are characterized by a periodic change of these parameter values from maximum to minimum with a frequency interval between adjacent maxima and minima defined by the approximate formula (4.7). Depending on the line length, this frequency range may be such that, within the nominal frequency band of the transmitter, there may even be several maximums and minimums of the HF path input impedance frequency characteristics and return loss.

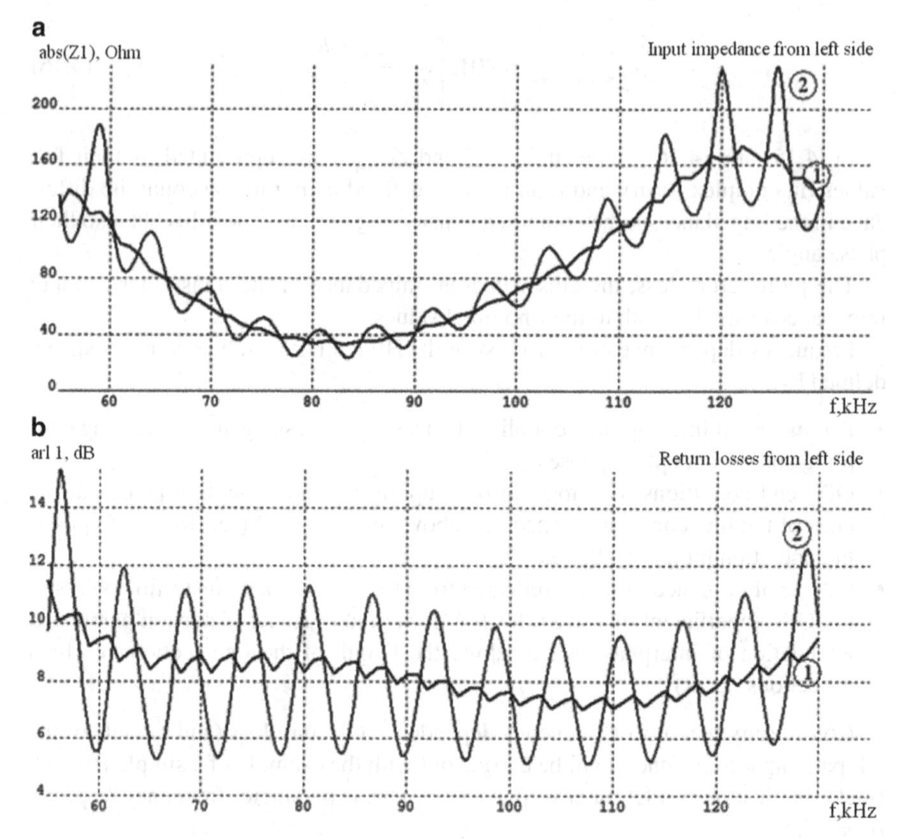

Fig. 4.37 Frequency dependence of the input impedance module (**a**) and return loss of the HF path (**b**) for short circuit at the ends of the overhead line. *Graph 1* OPL length is 300 km, *Graph 2* OPL length is 25 km

In this case, the question arises – how to ensure matching of the equipment and HF path, if for each transmitted spectrum frequency, matching conditions are different.

In addition, it should also be noted that frequency dependences of the parameters under consideration are unstable over time. They vary in some degrees when the substation to which the considered overhead line is connected switches (this causes changing end phase conditions for the considered overhead line and changing conditions of signal reflections from the end of the OPL).

We will illustrate the said above by an example of HF path along transposed 500 kV OPL with a length of 300 km (with almost absent reflected wave influence) and non-transposed 500 kV OPL with a length of 25 km and significant influence of reflected waves. In both cases, the path is built according to optimal coupling schemes using a typical coupling device (line matching unit) with 4650 pF capacitance and bandwidth of 54–130 and line trap having blocking impedance of 440 ohms.

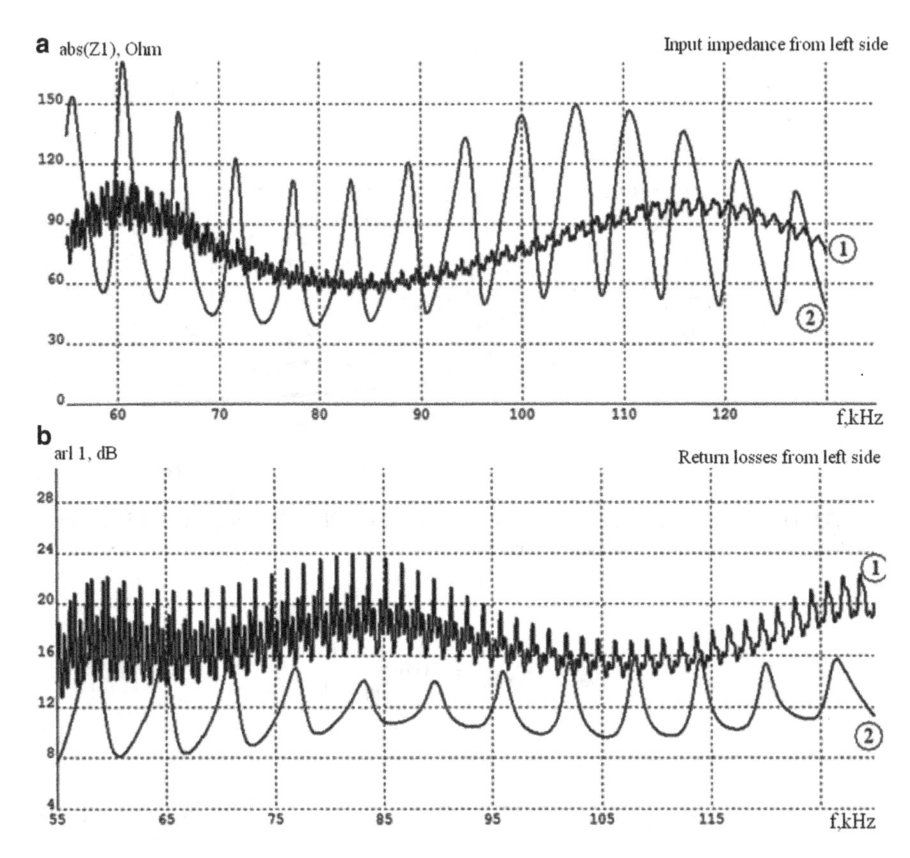

Fig. 4.38 Frequency dependence of the input impedance module (**a**) and return loss of the HF path (**b**) for open circuit at the ends of the overhead line. *Graph 1* OPL length is 300 km, *Graph 2* OPL length is 25 km

In addition, for each overhead line, we will consider two operation modes (the same for both ends): open and short circuit.

Calculation results for these paths are shown in Figs. 4.37 and 4.38.

In addition to these results, we note that, in the case of the overhead line with a length of 25 km, the input impedance phase angle changes with frequency and can reach a value of ±50°.

The figures clearly show features of frequency dependences of the HF path input impedance and return loss for the two edge cases.

Earlier, we have considered features of return loss when input impedance of the HF path is equal to its nominal impedance. When we discuss the degree of matching transmitter internal impedance (Z_{TX}) with input impedance of the HF path ($Z_{input.path}$), it is necessary to consider return loss between the transmitter and input impedance of the HF path. In this case, it is more convenient to write the formula for return loss in a slightly different form than the formula (4.15):

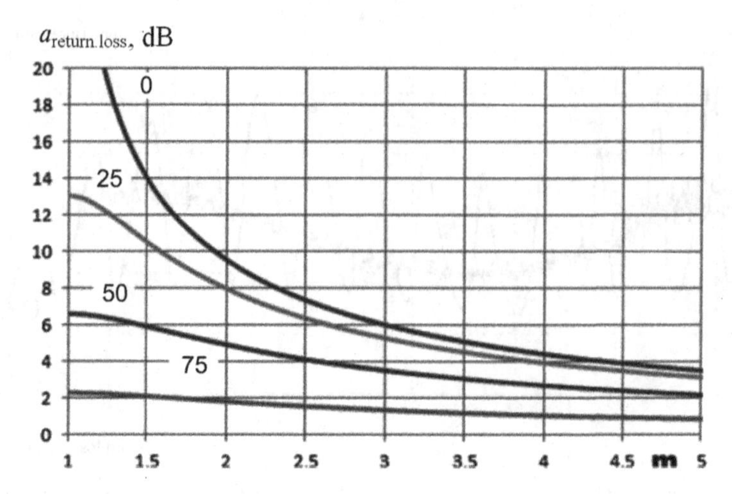

Fig. 4.39 Dependence *a*return.loss on module of ratio of the PLC transmitter and HF path total impedances at different phase angles of this ratio

$$a_{\text{return.loss.TX/path}} = 20\lg\left|\frac{m+1}{m-1}\right|$$

where:

$$m = Z_{\text{TX}} / Z_{\text{input.path}} \tag{4.17}$$

If we denote the phase angle of *m* coefficient as $\Delta\varphi$ (difference between phase angles of total impedances Z_{TX} and $Z_{\text{input.path}}$), then the dependence $a_{\text{return.loss.TX/path}} = \varphi(m)$ if module of $m \geq 1$ and $\Delta\varphi = 0^\circ$, 25°, 50°, and 75° has the form shown in Fig. 4.39. This figure is also valid if argument of $m_1 = 1/m$ ($m_1 \leq 1$).

As can be seen in Fig. 4.39, the best matching corresponds to the case when the angle $\Delta\varphi$ is zero (this corresponds to the case when both impedance Z_{TX} and impedance $Z_{\text{input.path}}$ are purely real). For identical modules *m*, the more the angle $\Delta\varphi$, the less matching degree.

Let us further consider the issue of transmitter matching necessity (desirability) to reduce undesirable consequences of the PLC equipment and the HF path mismatch.

Usually, matching is provided as linear winding of the line transformer of the PLC equipment transmitter has windings that allow obtaining different discrete values of the internal impedance module for this transmitter. Discreteness at this impedance selection (number of windings) varies for different manufacturers. Matching of the PLC equipment transmitter with the input impedance of the HF path is carried out by selecting such a tap, in which power of the signals transmitted

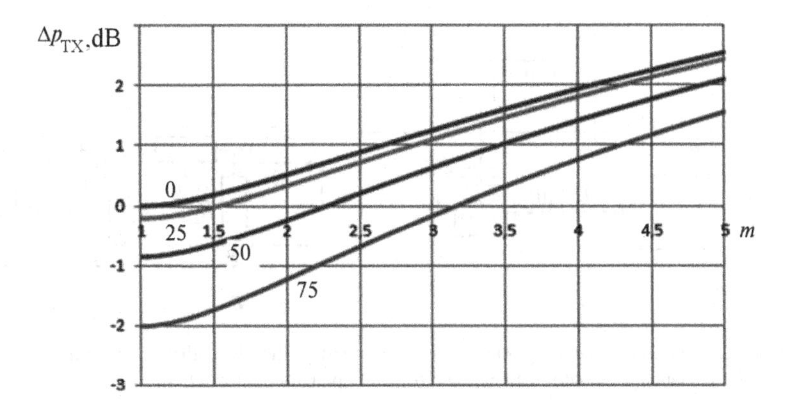

Fig. 4.40 Dependence of the transmission level on degree of the transmitter and the load mismatch

in the PLC channel, input to the HF path, is maximum possible. The system pilot HF signal can be used as the adjusting signal.

Let us consider changes of the HF signal power output by the transmitter with internal impedance of Z_{TX} to the load which is input impedance of the HF path $Z_{input.path}$ (further we will call it just "to load") at different degrees of the transmitter – load matching.

Let us discuss this for the two edge cases mentioned above, for behavior of return loss frequency characteristics and input impedance.

The first edge case (absence at the OPL beginning of waves reflected from its end). When considering this case, we assume that, within the transmitter nominal frequency band, change of the HF path input impedance can be ignored, considering it constant. Input impedance of the transmitter will be also considered constant.

Let us define dependence of the difference between transmission power for matched transmitter mode and mismatch level (Δp_{TX}) on degree of the transmitter input impedance correspondence to the load defined according to (4.17).

It is easy to obtain that this dependency has the form:

$$\Delta p_{TX} = 10\log\frac{P_{match}}{P} = 10\log\left|\frac{(1+m)^2}{4m}\right|,$$

where $P_{match,}$ signal power, W, delivered by the transmitter to the matched load; P, signal power, W, delivered by the transmitter to the mismatched load; and m, complex (in general case) coefficient defined according to (4.17).

When supposing that the phase angles of the total impedances Z_{TX} and $Z_{input.path}$ differ by the angle $\Delta\varphi$, then dependence $\Delta p_{TX} = \varphi(m)$ for module $m \geq 1$ and $\Delta\varphi = 0°$, 25°, 50°, and 75° has the form shown in Fig. 4.40.

This figure is also valid if argument of m is $1 = 1/m$ ($m_1 \leq 1$).

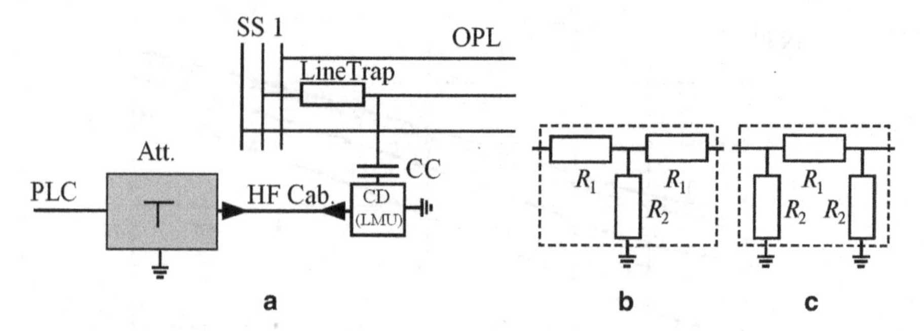

Fig. 4.41 Connecting a matching attenuator (at one end of the HF path): (**a**) general switching scheme; (**b**) T-shaped attenuator scheme; and (**c**) U-shaped attenuator scheme

For the variant when $m = 1$, the transmitter is fully matched with the load, and for lower (or larger) m values and at different phase angles of these impedances, the transmitter is not fully matched with the load.

As can be seen in Fig. 4.40, the largest transmitter power reduction when it is mismatched with the load corresponds to the case when angle $\Delta\varphi$ is zero (this corresponds to the case when both impedances Z_{TX} and $Z_{input.path}$ are purely real). But even in this case, at $m \leq 2$, the reduction of the transmission power in the case of mismatching is equal only to 0.5 dB, and in this case, it is practically not necessary to ensure matching of the transmitter.

The need for matching should only be considered when $m > 2.25$.

For the second edge case of the HF path along a short overhead line with a sufficient influence of repeatedly reflected waves, the use of transformer windings for matching to increase the transmission power in the DPLC channel cannot be recommended. This is due to the fact that the phase angle between the impedances Z_{TX} and $Z_{input.path}$ within the transmitter nominal frequency band can vary widely and matching in one part of the transmitter frequency band will lead to mismatch in another part of it that can lead to distortion of the transmitted DPLC HF signal spectrum with all possible consequences.

In addition, for a PLC channel along a short overhead line, the problem of reducing transmission power by 1 to 2 dB is not important.

In the case of short overhead lines, if better matching is required, for example, for improving the transmitter thermal mode, it can be recommended to install an attenuator at both ends of the HF path between the terminal PLC equipment and HF cable, as shown in Fig. 4.41.

Required attenuation of the additional attenuator installed at both ends of the HF path for matching the PLC equipment with the HF path (see Fig. 4.41) can be defined by the formula:

$$0_{att} = \frac{0_{return.loss.att} - 0_{return.loss.path}}{2} \tag{4.18}$$

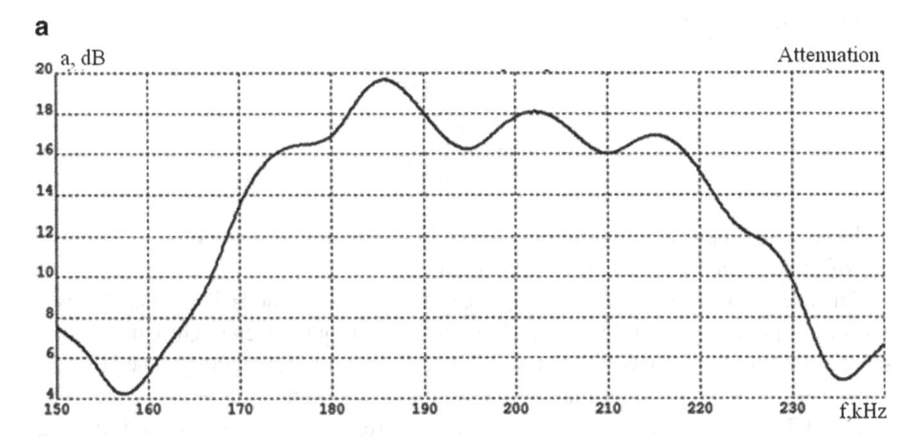

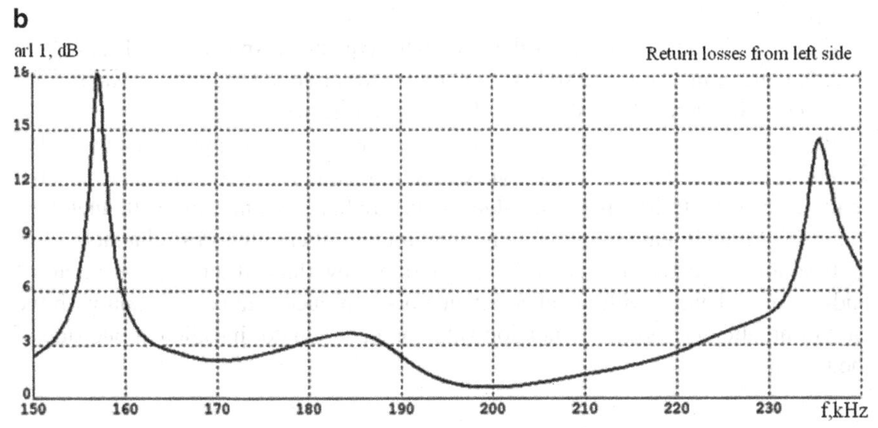

Fig. 4.42 Frequency characteristics of parameters for the HF path along COPL: (**a**) attenuation and (**b**) return loss

where $a_{\text{return.loss.att}}$, dB is the value of return loss required to ensure normal operating conditions of PLC equipment and $a_{\text{return.loss.path}}$, dB – return loss of the existing HF path.

If $a_{\text{return.loss.att}} = 12$ dB and, as the worst case scenario, $a_{\text{return.loss.path}} = 0$ for this edge case, and $a_{\text{att}} = 6$ dB is required.

Resistors R_1 and R_2 required for implementing an attenuator with attenuation a for the scheme shown in Fig. 4.41, dB can be defined by the formulas:

- T-scheme (Fig. 4.41b):

$$R_2 = \frac{Z_{\text{att}}}{sh\left(0{,}115a_{\text{att}}\right)}$$

$$R_1 = R_2\left(ch\left(0{,}115a_{\text{att}}\right)-1\right)$$

- U-scheme (Fig. 4.41c):

$$R_1 = Z_{\text{att}} \operatorname{sh}(0{,}115 a_{\text{att}})$$

$$R_2 = \frac{R_1}{\operatorname{ch}(a_{\text{att}}) - 1}$$

Here, Z_{att} is the characteristic impedance of attenuator, ohms (usually 75 ohms); a_{att}, dB, is the attenuator attenuation defined by (4.18).

The use of extensions for matching the transmitter with the HF path may be useful for HF paths along COPL in which attenuation frequency characteristics can be satisfactory but return loss frequency characteristics – not satisfactory. The HF path shown in Fig. 4.31, without matching circuits in the connection point of the OPL and the CPL, can be an example; the attenuation for this case is shown in Fig. 4.32 (Graph 2).

Figure 4.42 once again shows the line path frequency response from Fig. 4.28 in the frequency range 150 to 240 kHz, and return loss frequency characteristics at the left side of the path will be shown in the same frequency range.

It can be seen in Fig. 4.42 that, in the frequency range of approximately 180 to 210 kHz, the frequency response characteristic is suitable for creating channels, but return loss characteristic is not suitable for this. In this case, installing attenuators at the ends of the HF path allows solving the problem of creating PLC channels.

It should be taken into account that possible overlapped attenuation (channel budget) of a channel with installed attenuators is reduced (in comparison with the case of no attenuators) by attenuation value of an attenuator installed at one side of the HF path.

4.3 Characteristics of Disturbances in the HF Paths

Disturbances at the HF path output can be divided into wideband noise (due to the presence of high voltage and lightning overvoltage at the power line phase conductor) and narrowband interference (due to other PLC channels and radio stations).

Wideband noise, in turn, can be divided into continuous and transient.

Continuous wideband noise or corona noise should be considered when defining performance of PLC channels.

Temporary wideband noise or transient disturbance is usually so high that PLC channels fail during this noise presence.

Narrowband interference should be taken into account when selecting frequencies for PLC channels.

Let us consider parameters of the mentioned disturbances separately for each type in relation to their impact on the DPLC channels.

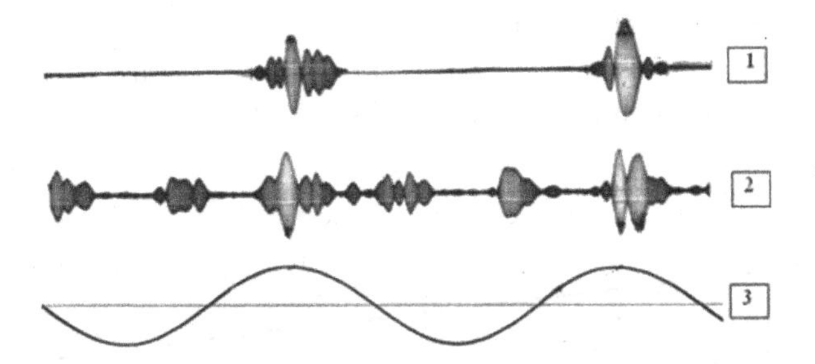

Fig. 4.43 Waveforms of corona noise in HF paths with different types of coupling to the OPL. (1) intra-phase path; (2) phase-to-earth path; and (3) 50 Hz phase voltage

4.3.1 Corona Noise

Under acting wideband noise channel performance is ensured by selecting minimum permissible reception level which provides required signal-to-noise ratio.

For lines with a voltage of 220 kV or higher, the main source of wideband noise at a normal line operation is corona on the line phase conductors. At bad weather conditions, corona can be considered as one of the main sources of wideband noise for 110 kV overhead line.

Corona is a phenomenon in which electric discharges appear in a certain volume of air near the surface of phase conductors due to a high electric field density. Noise from corona is caused almost exclusively by streamer discharges of the so-called "positive" corona which occurs in time intervals of approximately 6 ms, when voltage at this phase of the overhead line is close to positive maximum.

Since all three phases of the line produce corona, then, in general, three bursts of voltage occur at the output of the HF path during one period of industrial frequency. There are only two exceptions. The first case is a high-frequency path with a phase-to-phase outer connection to non-transposed lines with horizontal phase arrangement. At the output of this HF path, during the period of 50 Hz (20 ms), there are only two bursts of voltage (corona noise at center phase conductor in this path is almost absent). The second case is HF path with intra-phase coupling to isolated bundled-phase wires; there is only one burst of interference voltage during 20 ms at the output of this path (corona noise from another two non-coupling phases in this path is practically absent).

At each moment of time, the noise voltage due to corona is distributed according to the normal law. In this case, interference due to corona can be considered with enough confidence as interference of white noise type with a periodically changing root-mean-square (RMS) value.

In Fig. 4.43, waveforms of the corona noise voltage for various types of coupling to the overhead line are given.

For each of the phase angle value with a voltage vector of 50 Hz varying within one period of this voltage from 0 to 360 electric degrees, corona noise voltage is a random value distributed according to the normal law. Corona noise

Table 4.10 Nominal levels of corona noise

Normal line voltage, kV	110	220	330	500	750
Nominal corona noise level $p_{corona.50\%}$, dBm	−38	−28	−26	−20	−19

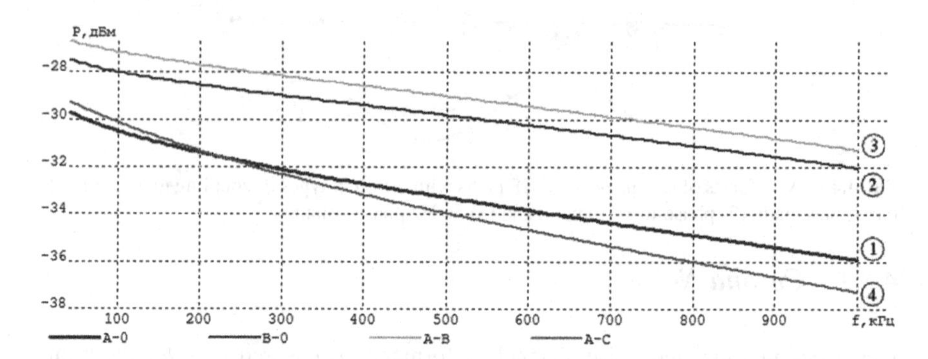

Fig. 4.44 Corona noise level for center phase-to-earth coupling (graph 1); outer phase-to-earth (graph 2); center phase-to-outer phase (graph 3); and outer phase-to-outer phase (graph 4)

root-mean-square voltage is a periodic function with the period equal to the period of 50 Hz voltage (360 electric degrees or 20 ms). Commonly, this function is described using not the absolute value of root-mean-square voltage but a value of the average root-mean-square noise voltage during this period. The type of this function depends on the line coupling scheme where noise is defined, on whether the line is transposed, on the weather conditions, and, finally, on the frequency range where noise is defined.

Let us consider the features of corona noise commonly used to characterize this noise during designing and operation of PLC channels.

We should note that the corona noise level is a random value which depends, at all other things being equal, on many external factors which change over time (line voltage, atmospheric pressure, temperature and humidity, precipitation intensity, location along the OPL route, etc.).

Precipitations in the form of rain and snow and air pollution mainly affect the level of corona noise. Taking into account that the area currently subject to heavy precipitation is usually not very large, the highest noise is generated under heavy precipitation in the line section adjacent to the substation where the noise is defined (of course, the longer the OPL section temporary subject to precipitation, the higher noise level is expected).

The difference between the levels of noise defined on the same line without precipitation and mist (let us call it fair weather) and with heavy precipitation may reach 20 to 30 dB. The difference between noise levels measured under seemingly identical conditions can reach up to 5 to 10 dB.

Let us consider the features used to describe corona noise when calculating noise immunity of PLC channels.

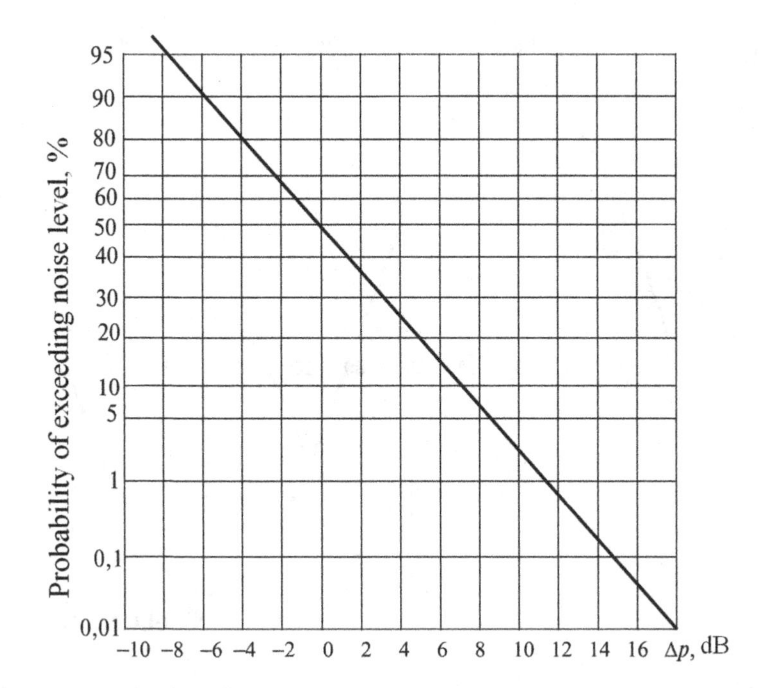

Fig. 4.45 Graph for determining the probability of exceeding noise level above the level of ($p_{noise.50\%}$ + Δp)

Nominal Corona Noise Level

Nominal corona noise level $p_{noise.}$ or $p_{noise.50\%}$ is defined as a level of average root-mean-square noise voltage with a 50 percent probability of exceeding.

Table 4.10 provides for ratios of nominal corona noise levels for phase-to-earth coupling to lines of different voltage classes typically used for PLC channel design and presented in [12]. These ratios relate to 100 kHz frequency in a band of 1 kHz with a 50 percent probability of level increase ($p_{noise.50\%}$). The same ratios can also be used for phase-to-phase coupling schemes.

The corona noise level depends on the bandwidth Δf in which the noise is defined. When data on the noise level defined in the frequency band Δf_1 are available (based on measurements or calculation results), then this level can be recalculated to another frequency band Δf_2 using the formula:

$$p_{noise\Delta f_2} = p_{noise\Delta f_1} + 10\log\left(\Delta f_2 / \Delta f_1\right)$$

Thus, for instance, the noise level in 4 kHz band is 6 dB higher than the one in Table 4.10.

Data in Table 4.10 present generalized results of calculation carried out in the WinNoise program. Software algorithm is based on an accurate calculation method described in [14].

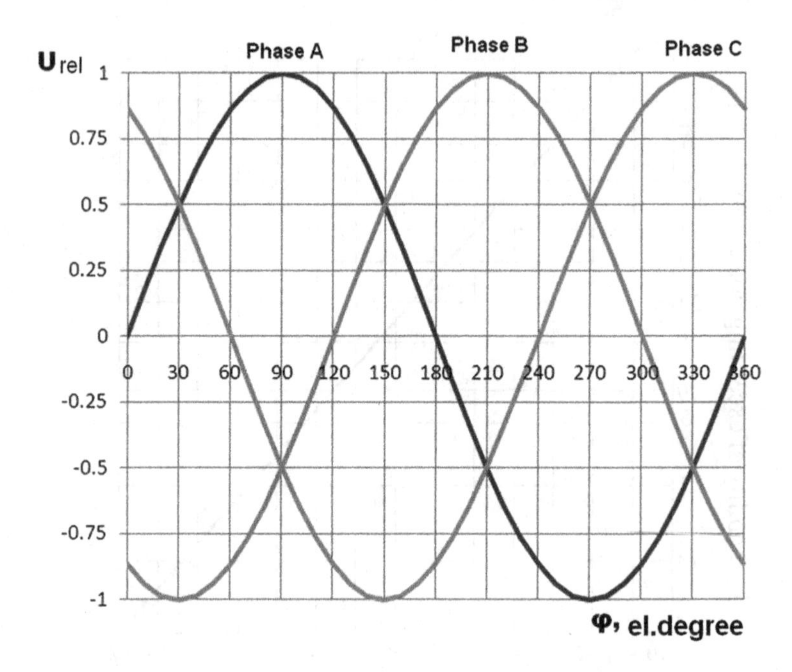

Fig. 4.46 Dependence of the relative voltage of phases A, B, and C on the phase angle φ. Graph 1 for phase A, Graph 2 for phase B, and Graph 3 for phase C

Figure 4.44 provides for results of corona noise calculation with a 50 percent probability for 220 kV OPL with horizontal phase conductor arrangement (OPL length of 50 km).

Integral Distribution Function of Noise

Integral distribution function of noise as a random value $P(p_{noise.} \geq p_{noise.50\%} + \Delta p))$ shows a probability for exceeding the nominal noise level $p_{noise.50\%}$ by a given value Δp. This function is shown on Fig. 4.45.

Dependence $U_{noise.rms} = f(\varphi)$.

To calculate noise immunity of PLC channels from corona with a sufficient probability level, one can assume that white noise with the dependence of root-mean-square noise voltage on the phase angle φ (electric degree) of 50 Hz ($U_{noise.rms} = f(\varphi)$) is a periodic function with a period equal to the period of 50 Hz voltage (360 electric degrees or 20 ms).

In practical use of this dependence, the main factor is the ratio of the maximum root-mean-square values of voltage in each of the bursts of noise. Thus, this dependence is calculated for relative root-mean-square voltage, $U_{rel} = U_{noise.rms}/U_{noise.ave\ rms}$ as: $U_{rel} = f(\varphi)$.

In addition, one should define the impact of OPL coupling and precipitation on the OPL route for this dependence.

As it was mentioned, $U_{rel} = f(\varphi)$ dependence can be achieved for the period of 50 Hz in one to three bursts, each caused by a positive corona on the relevant OPL phase.

The location of the maximum root-mean-square noise voltage in each of the bursts can be defined considering that voltage vectors of A, B, and C phases are shifted toward one another at an angle of $\varphi = 120$ electric degrees (or 20/3 ms time interval).

Dependence of the relevant voltage of all three phases (Y axis value corresponds to the phase voltage amplitude in relative units) on the phase angle φ (electric degree) is shown on Fig. 4.46. It reveals that if phase A voltage phase angle is taken as a reference, the burst of root-mean-square noise voltage caused by a corona at phase A is placed near a 90 degree angle, at phase B – near a 210 degree angle, and at phase C – near a 330 degree angle.

Now let us consider the nature of dependence $U_{rel} = f(\varphi)$ for different coupling schemes and various weather conditions at an OPL route. We will use the previously

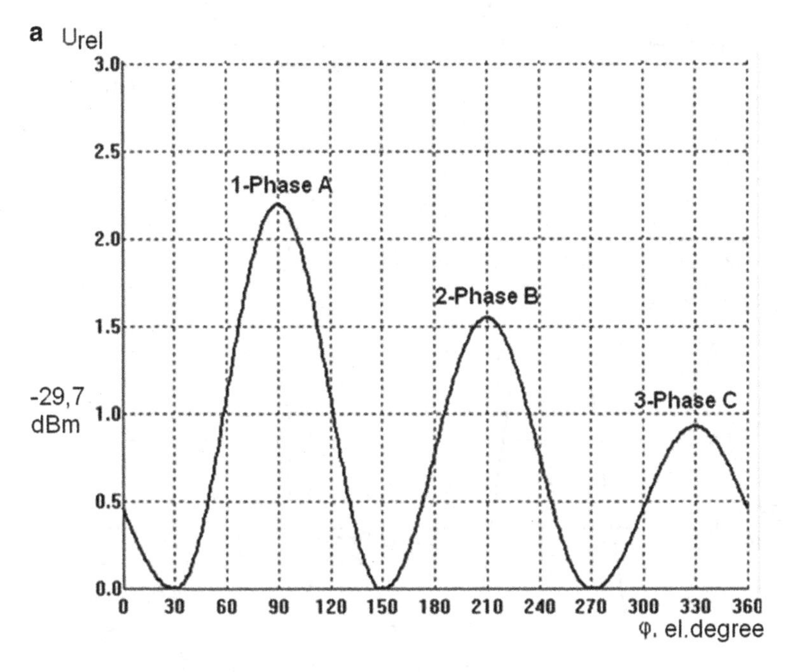

Fig. 4.47 Dependences of the relative RMS corona noise voltage on phase angle of the industrial frequency voltage with different coupling schemes and various weather conditions. The absolute level provided in the (**a**–**d**) figures' description is the power level calculated based on the average root-mean-square noise voltage in the considered case. (**a**) Phase A-to-earth; fair weather; ($U_{rel} = 1$ corresponds to power level of −29.7 dBm). (**b**) Phase B-to-earth; fair weather; ($U_{rel} = 1$ corresponds to power level of −27.5) dBm). (**c**) Phase A-to-earth; heavy precipitation near the substation; ($U_{rel} = 1$ corresponds –power level of −23.0 dBm). (**d**) Phase A-to-earth; heavy precipitation near the substation; ($U_{rel} = 1$ corresponds to level of power of −20.0 dBm)

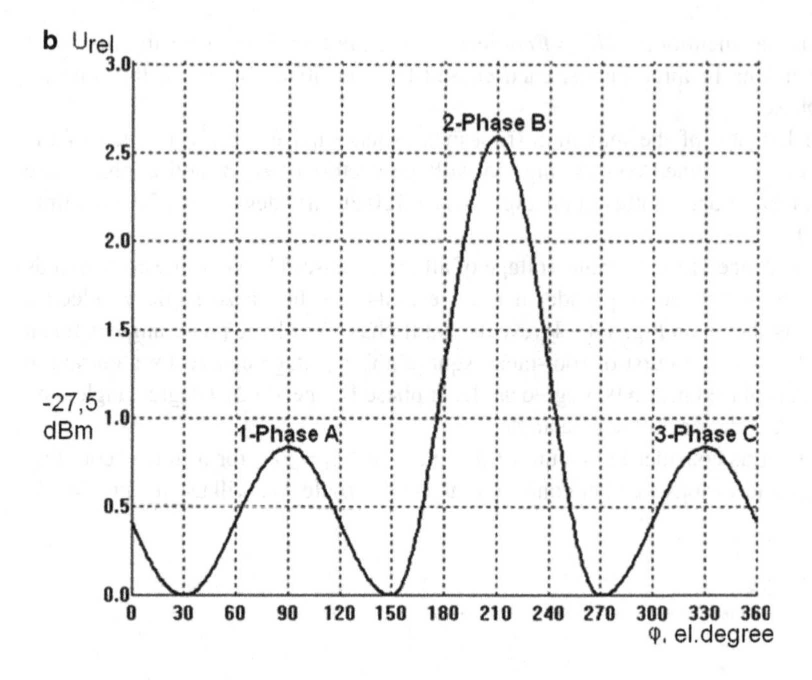

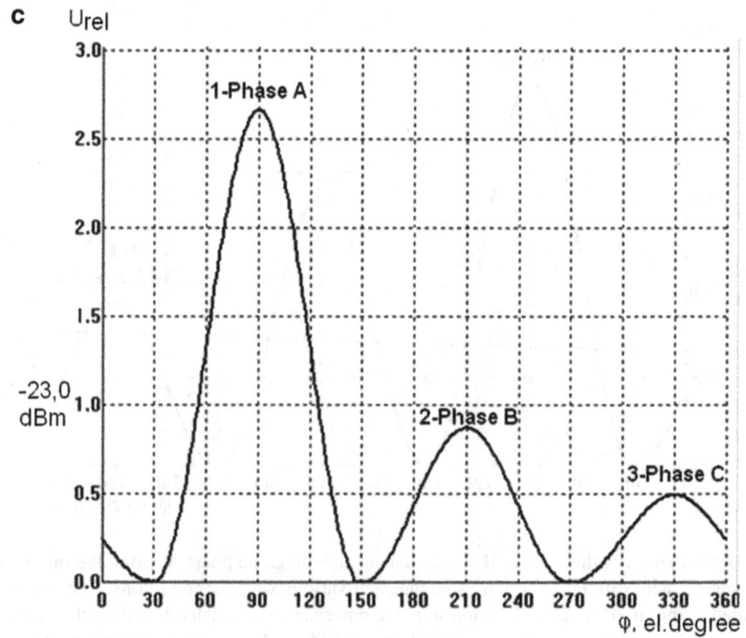

Fig. 4.47 (continued)

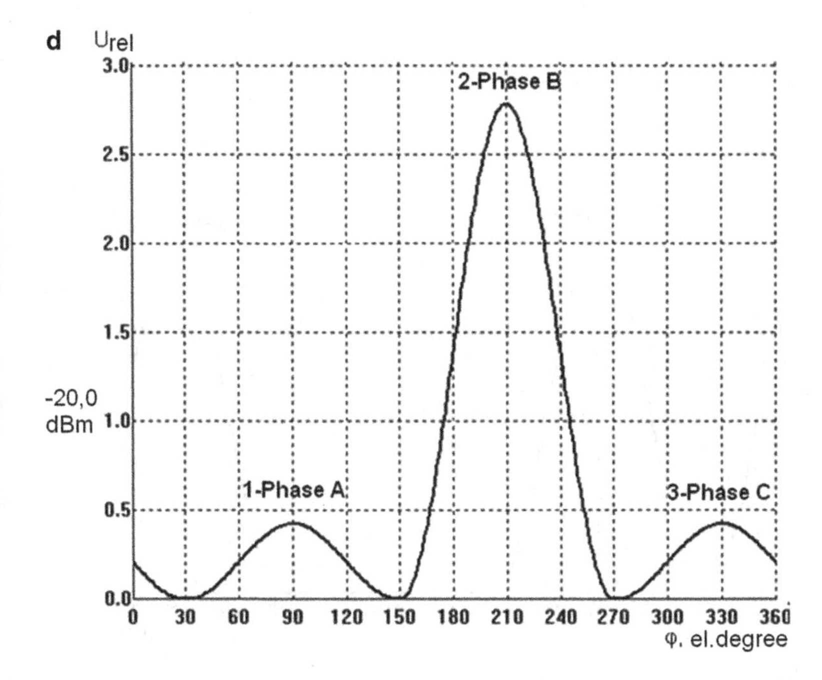

Fig. 4.47 (continued)

mentioned example of untransposed 220 kV OPL with horizontal phase arrangement. OPL length is 50 km. The frequency of noise equals 40 kHz. Noise is defined in 1 kHz frequency band.

Phase-to-Earth Coupling Schemes

At first, let us consider the dependence of $U_{rel} = f(\varphi)$ under fair weather conditions over the entire OPL path length. In this case, root-mean-square corona noise voltage in the maximum noise burst corresponding to the positive voltage maximum of 50 Hz at "own" phase can both differ greatly and be commensurate of the maximum root-mean-square voltage of burst corresponding to a positive maximum voltage of 50 Hz on "other" phases.

It is clearly shown in Fig. 4.47a, b, with the dependences $U_{rel} = f(\varphi)$ under the mentioned coupling conditions of outer phase A-to-earth and center phase B-to-earth (the figures imply that the phase angle is set for phase A when considering the coupling scheme for any phase). These figures also provide for average root-mean-square noise voltage levels (they can be obtained from Fig. 4.42).

Now let us consider the same dependences under adverse weather conditions (heavy precipitation) at the OPL section 5 km long adjacent to the substation where noise is defined. Let us assume that the function of noise generation (characterizing the power of streamer corona sources) at this section increases related to this function by 15 dB at the rest of OPL section.

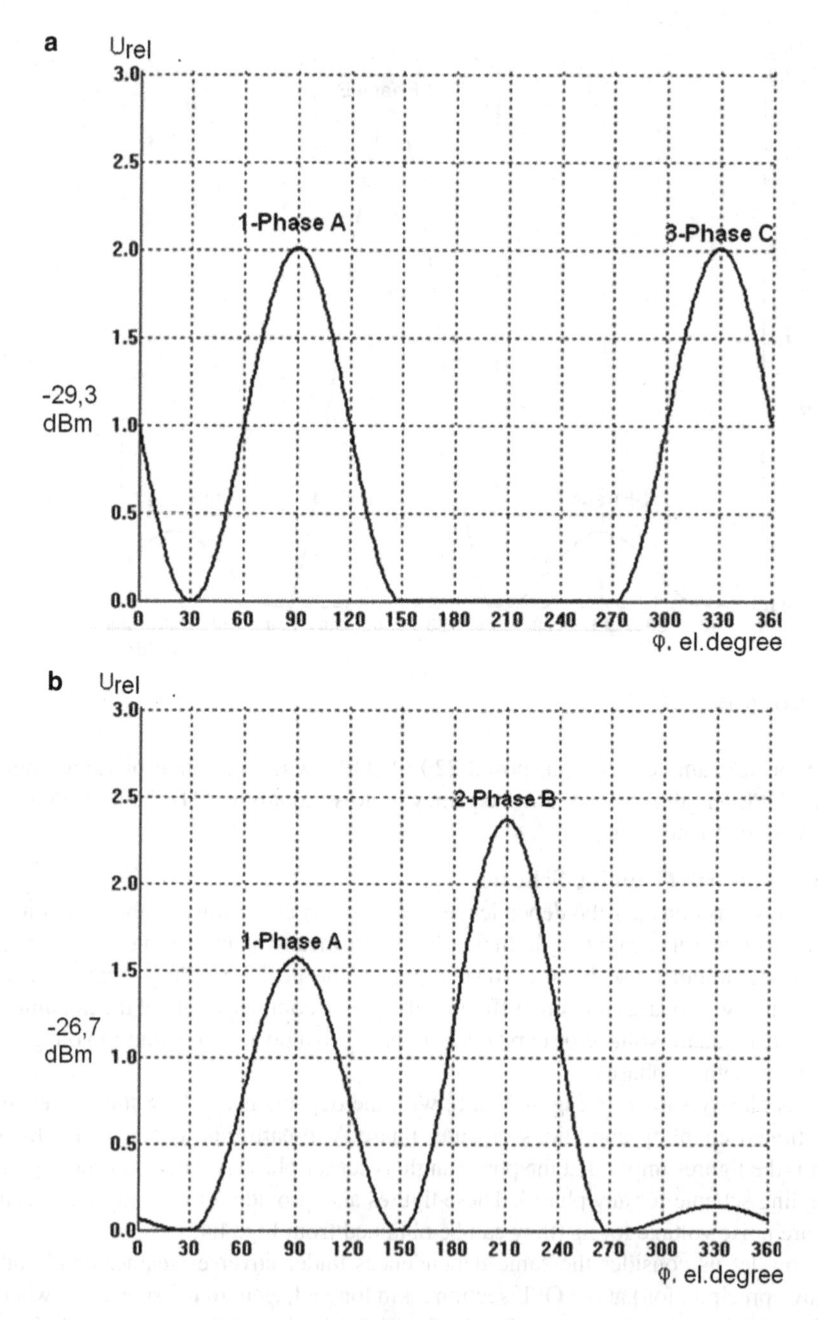

Fig. 4.48 Dependences of the relative RMS corona noise voltage on phase angle of the industrial frequency voltage with different coupling schemes and various weather conditions. The absolute level provided in the (**a**–**d**) figures' description is the power level calculated based on the average root-mean-square noise voltage in the considered case. (**a**). Phase A-to-phase C; fair weather; ($U_{rel} = 1$ corresponds to power level of −29.3 dBm). (**b**). Phase A-to-phase B; fair weather; ($U_{rel} = 1$ corresponds to power level of −26.7 dBm). (**c**). Phase A-to-phase C; heavy precipitation near the substation; ($U_{rel} = 1$ corresponds to power level of −22.6 dBm). (**d**). Phase A to phase B; heavy precipitation near the substation; ($U_{rel} = 1$ corresponds to power level of −20.2 dBm)

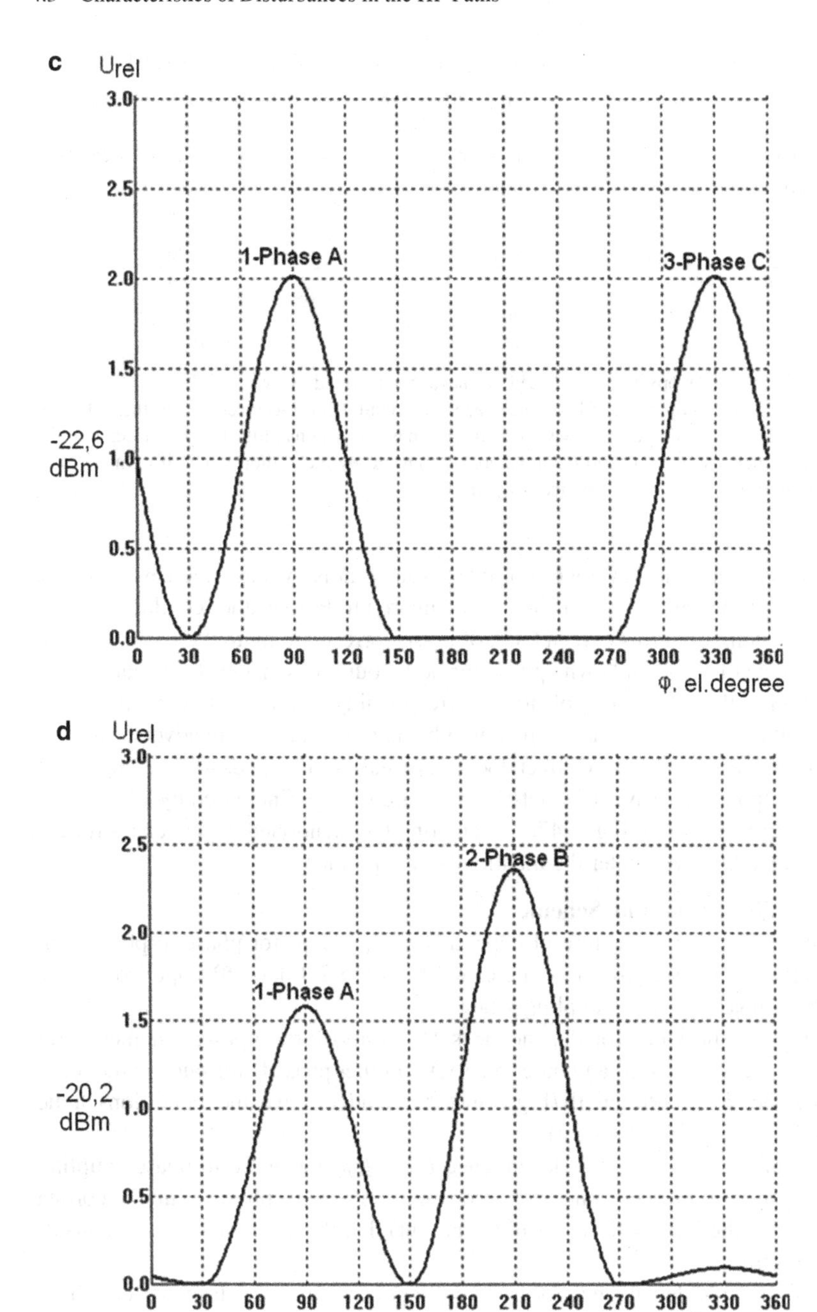

Fig. 4.48 (continued)

Table 4.11 Corona noise levels for alternating current OPL according to IEC 60488-1

OPL voltage, kV	Corona noise levels [dBm] in a 4 kHz band for different weather conditions	
	Fair weather	Adverse weather
Up to 110	−50	−30
150	−45	−25
220	−40	−20
300	−35	−15
400	−30	−10
800	−20	0

In this case not only average root-mean-square noise voltage increases, but the form of dependency $U_{rel} = f(\varphi)$ changes compared to fair weather conditions at the entire OPL length. In this case, a burst of noise corresponding to the positive maximum voltage of 50 Hz at "own" phase for both coupling schemes turns out significantly higher than the bursts of noise corresponding to the positive maximum of 50 Hz voltage on "other" phases. It should be mentioned that if an adverse weather area covers the entire line, the effect does not reveal itself. It leads to changing of the level corresponding to average root-mean-square voltage (increased by 15 dB).

It is clearly shown in Fig. 4.47c, d depicting the mentioned dependences for the same 220 kV OPL yet under the mentioned precipitation.

Phase-to-Phase Coupling Scheme

Now let us consider the nature of dependency $U_{rel} = f(\varphi)$ for phase-to-phase coupling scheme. We will use the same example of the 220 kV OPL applied for the analysis of phase-to-earth coupling scheme.

Figure 4.48 provides for dependences $U_{rel} = f(\varphi)$ for coupling schemes outer phase A-to-center phase B and outer phase A-to-outer phase C for fair weather conditions at the entire OPL length (Fig. 4.48a, b) for adverse weather conditions at the section 5 km long (Fig. 4.48c, d).

As shown in Fig. 4.48, the dependence $U_{rel} = f(\varphi)$ for phase-to-phase coupling practically reveals only two bursts of corona noise on "own" phases. Burst of corona noise on "another" phase is so small (or absent) that the third burst can be considered negligible.

In addition, the dependence form $U_{rel} = f(\varphi)$ in case of phase-to-phase coupling, as opposed to phase-to-earth coupling, is practically not subject to weather conditions at the OPL route (of course, the absolute root-mean-square noise voltage value is impacted by the weather).

We will not include the analysis of reasons for the above patterns of dependences $U_{rel} = f(\varphi)$ for phase-to-earth and phase-to-phase couplings. We will only mention

Table 4.12 Comparison of the results of corona noise calculation

Corona noise level calculation, dBm using generation function according to				
According to [6]			CIGRE [16, 17]	
$P_{noise.50\%}$, dBm (numerator is taken from WinNoise and denominator is taken from Table 4.10)	Δp, dB (exceeding probability of 5%) according to Fig. 4.45	$P_{noise.5\%}$	$P_{noise50\%}$	$P_{noise5\%}$
(− 27.5)/(−28.0)	8.4	(−19.1)	(−28.9)	(−22.1)

that these patterns are caused by different significances for noise formation in the share caused by distributing the noise from the sources located on the line to the end of the zero-mode path (all phases-to-earth mode).

It appears to be interesting to compare the data on the typical values of the corona noise levels on the lines of different voltage classes provided in [12] and with the data provided in IEC 60488-1 standard (Table 15). This table is given as Table 4.11 [15].

We should note that provided values may vary greatly due to difference in the structure of OPLs causing the differences of electric field voltage at the surface of phase conductors.

At that, such a comparison is beyond doubt necessary considering that in [12] the noise level is calculated and regulated for a 1 kHz band and in IEC 60488-1 standard it is calculated and regulated for a 4 kHz band.

However, this comparison is impossible in practical terms. A reliable comparison is rendered impossible mainly due to the following reasons:

In [12] (Table 4.10 and Fig. 4.45), a nominal noise level $p_{noise.50\%}$ (with a 50 percent probability of exceeding), and a correction to value $p_{noise.50\%}$, is used, allowing to define the level of noise to be exceeded with the set probability.

In Table 4.11 (IEC 60488-1, Table 15), noise level is provided for two weather variations (fair weather and adverse weather). However, in these weather conditions, the probability of exceeding the mentioned noise levels is not defined, which renders their comparison with regulations in [12] impossible.

Design features impacting the corona noise level (diameter of the used phase conductors, number of wires in bundled-phase, etc.) are different for typical OPLs in different countries.

As evident from notes to Table 15, data on corona noise level in IEC 60488-1 are illustrative and are not intended be used in calculations.

In this regard, the best way to compare the method of corona noise calculation is by using the WinNoise software and generation function defined in [6] and the method recommended by CIGRE [16, 17] (this generation function defines the corona noise level when the rest of the conditions are the same).

We will make a comparison for the 220 kV OPL mentioned above, for center phase B-to-earth coupling (at a 100 kHz frequency and 1 kHz band).

In calculating, let us assume that the corona noise excitation function using CIGRE [16, 17] is defined for two cases. These cases correspond to a 50% and 95%

Table 4.13 Parameters of transient disturbances caused by various sources

Noise source	Level of p_{noise}, dB_v	Average pulse number per second	disturbance duration, ms
Enabling and disabling sections of buses and equipment using a disconnector	+25	2–3 at the beginning of switching on or at the end of switching off; 100–1000 the rest of the time	500 to 5000
Enabling and disabling the line using a circuit breaker	+20	1000–2000	5 to 20
Beginning of steady arc discharge	+25	1000–2000	2 to 5
Steady arc discharge	−10	100–300	–
Disabling short current using a circuit breaker	+25	1000–2000	10 to 30
Lightning discharges	+25	2–3 (up to 40)	≤1000

Notes:
(1) The disturbance level is given for 4 kHz frequency band
(2) The disturbance level is defined as 20 log(U_{noise}/0,775), where U_{noise} is the effective value of the sinusoidal voltage amplitude of which is equal to the pulse amplitude

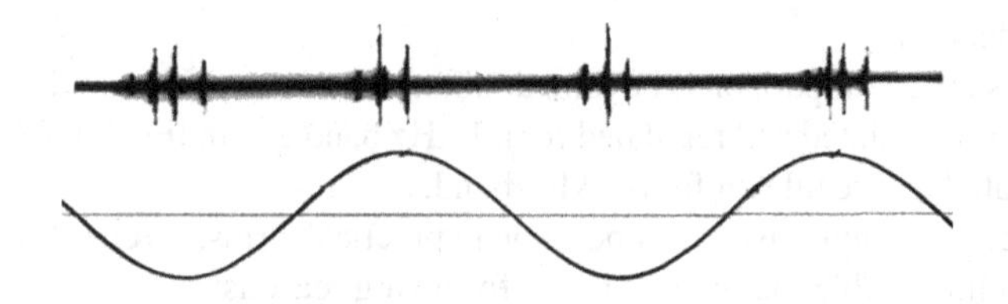

Fig. 4.49 Waveforms of the disturbances due to disruption of spark gaps on groundwire

probability, respectively, that the generation function will not exceed the value defined in this case.

In terms accepted in [12], it corresponds to defining the level of noise $p_{50\%}$ and $p_{5\%}$, with indices of 50% and 5% corresponding to the probability of level increase.

We will put the results of corona noise calculation according to WinNoise with excitation function according to method [6] and recommended by CIGRE [16, 17] in the Table 4.12.

Thus, it is evident that corona noise levels defined according to the method used in [6] almost coincide with the interference defined using CIGRE recommendations.

4.3.2 Transient Disturbances

Transient wideband noise is impulsive and is caused by lightning discharges during a thunderstorm and the electric disruption of air gaps which occur:

- In the place of the short circuit in the electric grid.
- Between blades of disconnectors and contacts of circuit breakers when switching on or off these high-voltage switching devices.
- When the spark gap which protects the steel groundwire insulation suspended on the overhead line is disrupted. This happens when the earthing of this cable is disrupted for some reason.

Information on the transient disturbance's parameters can be found in Table 4.13.

Similar to corona noise, transient disturbances are wideband, i.e., its spectrum covers the entire frequency range allocated for PLC communications. Maximum level of these disturbances is very high (much higher than the level of noise due to corona), and it is impossible to tune out of it and maintain operation of the PLC channels during acting of these interferences.

However, due to the small duration of these disturbances, degradation of the channel performance during this time can be not considered. The exception is noise due to disruption of spark gaps which protect insulation of steel groundwire which can last indefinitely and permanently disrupt operation of PLC channels. These interferences occur in cases when, for some reason, groundwire earthing failed and high-voltage relative to the earth is induced on it through capacitive connections with the phase conductors. To ensure normal operation of PLC channels in this situation, the only measure is to restore the earth groundwire. The feature of this type of disturbance which can be used to identify the source of fault is that the noise occurs twice during a period of 50 Hz near maximum of positive and negative half-waves of the voltage in the groundwire. The typical characteristic of this noise at a voltage of 50 Hz for the affected groundwire is shown in Fig. 4.49. Note: It is necessary to keep in mind that the voltage vector phase angle in the cable does not coincide with the phase angle of the vector of any phase voltage.

4.3.3 Narrowband Interferences

Narrowband interference caused by signals from radio stations operating in the area of the overhead line and all PLC channels operating in the electric grid is considered continuous. Performance of PLC channels in the presence of these interferences is provided by an appropriate selection of frequencies for the channel in question. This selection should ensure compliance with the requirements for mutual interference between the new channel and other channels operating in the common electric grid.

Signals in channels along some overhead line are distributed over the electric grid far beyond the borders of this OPL. For any PLC channel, signals transmitted by other channels are interference, and therefore their levels and frequencies should be considered when analyzing performance and planning of the PLC channels.

Propagation of signals in a grid of the same voltage occurs because all lines of this grid are conductively connected to each other through the busbars of the substation between which these lines are installed. Between networks of different voltages, propagation occurs by means of capacitive coupling between different voltage windings in power transformers installed at the substation. In addition, different parts of the network can be "connected" with each other by means of a high-frequency field in the points where they come close to each other or intersect different lines due to electromagnetic coupling in the point of proximity (intersection).

This propagation is limited by the natural attenuation of signals along the lines where these signals are propagated, by attenuation caused by shunting action of the substation, and by line traps included in the lines. In relation to line traps, it should be remembered that they are not always included in all phases and, in addition, they limit only those signals whose frequencies are located within the line trap blocking frequency range.

References

1. Carson, J.R.: Wave propagation in overhead wires with ground return. Bell Syst. Technol. J. **5**(4), 539–554 (1926)
2. Carson, J.R., Hoyt, R.S.: Propagation of periodic currents over a system of parallel wires. Bell Syst. Technol. J. **6**(3), 495–545 (1927)
3. Fallou, J.: Propagation des courants de haute frequency poplyphases le long les lines aḗrienne de transport l'ḗnergie. Bulletin de la Sosiete Francaise des Electriciens, Aout, (1932)
4. Kafeiva, K.I.: Corona Disturbances on the Conductors of Transmission Lines (in Russian). GEI, Moscow (1963)
5. Kostenko, M.V., Shkarin, Y.P.: Calculation of HF path parameters along power transmission lines. Newsteller of USSR Academia of Sciences, Power and Transport, Moscow, №1 (1967)
6. Kostenko, M.V., Perelman, L.S., Shkarin, Y.P.: Wave processes and electrical disturbances in multi conductor high voltage electrical power transmission lines (in Russian). Energy, Moscow (1973)
7. Kostenko, M.V., Kadomskaya, K.P., Levinstein, M.L., Efremov, I.A.: Overvoltage and overvoltage protection for overhead and cable high voltage power transmission lines (in Russian). Science, St. Petersburg (1988)
8. Wedepohl, L.M.: Application of matrix methods to the solution of travelling-wave phenomena in polyphase systems. Proc. IEE. **110**, 2200–2212 (1963)
9. Wedepohl, L.M.: Wave propagation in nonhomogeneous multiconductor systems using the concept of natural modes. Proc. IEE. **113**, 622–626 (1966)
10. Wedepohl, L.M., Wasley, R.G.: Propagation of carrier signals in homogeneous, nonhomogeneous and mixed multiconductor systems. Proc. IEE. **115**, 179–186 (1968)
11. Shkarin, Y.P.: Measurements in high voltage PLC communications. WinTrakt and WinNoise software for computation of high frequency characteristics of the high voltage power lines and corona noise (in Russian). Analytic, Moscow, (2015). http://www.analytic.ru/articles/art450.pdf. Accessed 01 June 2020

12. Standard of Organization: Guidelines for Selection of Frequencies of High-Frequency Channels Along 35, 110, 220, 330, 500 and 750 kV Electric Power Transmission Lines. Federal Grid Company of the Unified Energy System of Russia, Moscow (2010)
13. Bicford, J.P., Mulieux, M., Reed, J.R.: Computation of Power System Transients. Peter Peregrimus Ltd, London (1976)
14. Shkarin, Y.P.: Method of calculation of HF disturbances due to corona noise on the conductors of transmission lines. Electr. J. **3**(3), (1982)
15. IEC 62488–1 Power line communication systems for power utility applications – Part I. Planning of analogue and digital power line carrier systems operating over EHV/HV/MV electricity grids. International Standard, IEC (2011)
16. SC36 CIGRE Interferences produced by corona effect of electric systems (Description of phenomena and practical guide for calculation. CIGRE WG36.01 (1966)
17. SC36 CIGRE Interferences produced by corona effect of electric systems. CIGRE WG36.01 (1974)

Chapter 5
DPLC Channel Design Issues

5.1 Main Tasks of the DPLC Channel Designing

When designing DPLC channels, the following issues should be addressed:

1. Collection of data on number, types, and properties of information signals to be transmitted via the DPLC channel
2. Selection of DPLC equipment depending on properties of the set of transmitted signals, such as IP traffic
3. Defining necessary parameters of DPLC equipment modems. These include information bit rate, bandwidth, transmission delay, and so on
4. Collection of data on power lines where the DPLC channel or channels are planned to be created
5. Selection of optimal scheme for coupling to power lines
6. Analysis of the climate map for power line location area
7. Calculations of HF path attenuation without IHRD
8. Calculation of noise level in the band of the DPLC equipment receiver
9. Calculation of IHRD effect on attenuation and noise of the HF path
10. Calculation of HF path parameters under various climatic and operating conditions of the power line
11. Calculation of the PLC channel attenuation and selection of the PLC channel upper frequency limit considering IHRD, to ensure specified reliability of the PLC channel

Normative and technical documents used in the designing DPLC channels are listed in Chap. 1; also IEC 60870-5-101 [1] and IEC 60870-5-104 [2] standards should be followed.

When solving DPLC channel design tasks, we will constantly refer to the information provided earlier in the previous chapters.

A. G. Merkulov et al., *High Voltage Digital Power Line Carrier Channels*,
https://doi.org/10.1007/978-3-030-58365-1_5

5.2 Types and Characteristics of DPLC Information Traffic

Every industry has its own requirements for transmitting process control information. In the eclectic power industry, several types of information with different priorities are transmitted between substations and dispatch center. In Table 5.1, data on the type of transmitted information, its priority, and type of the electric grid units interaction during the information transmission are given.

Let us analyze features of each signal:

- Dispatcher voice channels (DispVoice) are used for operational dispatch management of electric grid.
- Technological voice channels (TechVoice) are used by the company's departments for various purpose works.

Table 5.1 Characteristics of information signals transmitted via digital DPLC channels

Information signal/designation	Priority	Characteristics
Dispatching voice communication/(Disp.Voice)	High	Is used by dispatcher services for voice communications between DC and SS
		Interaction: DC ←→ SS[a]
		Force connection/dial connection. Protocol stack for VoIP: IP/UDP/RTP/speech codec
Data transmission from telecontrol system (TC)	High	Is used for transmitting data from TC RTU to server in dispatch center
		Interaction: DC ←→ SS[a]
		Operation mode: polling/sporadic transmission
		Protocol stack: IP/TCP/IEC-60870-5-104, variable data transfer rate
		Protocol stack: FT1.2/IEC-60870-5-101, constant data transmission rate
Data transmission from power meters (PM)	Medium	Is used for transmitting data from substation power metering counters to server in dispatch center
		Interaction: DC ←→ SS[a]
		Mode of operation: time interval polling
		Protocol stack: IP/TCP/IEC-60870-5-104
		Protocol stack: V.24/application layer protocol
Technological voice communication/(Tech.Voice)	Normal	Is used by substation staff for voice communication between substations
		Interaction: DC ←→ SS, SS ←→ SS
		Dial connection
		Protocol stack for VoIP: IP/UDP/RTP/voice codec
Other data services: e-mail, SNMP management, Internet/file transfer (AUX)	Low	At bandwidth resources availability
		Protocol stack: IP/TCP – guaranteed delivery; IP/UDP – non-guaranteed delivery

[a]*DC* dispatch center, *SS* substation

Two types of low bit rate speech codecs, G.729 and G.723.1, are mainly used for speech transmission for both dispatcher and technological communication channels. Voice signal is transmitted via the channel with the addition of small service headers that increases required bit rate by not more than 10% (see Chap. 3). When using voice over IP technology for DPLC channels, on the contrary, methods decreasing volume of IP packets headers which was also discussed in Chap. 3 are used.

Signals of telecontrol system consist of signals of teleindication (TI), telemetering (TM), and telecommand (TC).

Information about the switching equipment state – circuit breakers and disconnectors and other substation equipment – is transmitted by means of TI signals.

These are discrete signals which are changed relatively rarely: when switching equipment is switched on or off and so on.

Telemetering signals transmit information about the values of current I, voltage U, active P, and reactive Q power and frequency of each line. TM data transmission is initialized when I and U values change by a specified relative value, for example, 1%, or within specified time intervals.

The number of telemetering and teleindication signals depends on the substation electric scheme and number of incoming power lines. Characteristics and requirements for the telecontrol operation are given in [2]. Data are transmitted using IEC 60870-5-101 protocols when asynchronous transmission FT1.2 protocol via RS-232 interface is applied or IEC 60870-5-104 [1] when the data are transmitted using IP network protocols.

Let us index the telecontrol signals as TC. In Table 5.2, characteristics of telemetering and teleindication signals are given [2].

Data from power meters are used in compiling the energy balance and calculating power exchange. Data are transmitted by means of IEC 60870-5-104 protocol using IP network protocols or by means of V.24 asynchronous protocol and RS-232 interface.

The server performs polling the power meters cyclically at a specified time interval, for example, 15 or 30 minutes. The volume of data transmitted from a single metering device depends on its type and is approximately 200 bytes.

Let us index signals of power metering system of commercial electric power metering as PM.

Auxiliary data transmission services do not affect the process of the electric grid management; they include e-mail, access to the local network of the power utility, Internet access, and so on. Traffic from these services is initially low priority and not critical to transmission delay. Therefore, the bandwidth resource of network segments for this data varies depending on high-priority signal traffic of the channel.

Table 5.2 Characteristics of TM and TI signals

Signal	Volume of information, bytes	Characteristics
Telemetering	15/8	With time stamp/without time stamp
Teleindication	11/4	With time stamp/without time stamp

Let us index all auxiliary services signals as AUX. Currently, AUX services use IP network protocols to transmit data. Data can be transmitted in the guaranteed and non-guaranteed delivery modes.

During periods of exposure to adverse weather factors, DPLC equipment adjusts their bit rate, and multiplexers disable low-priority signals. This procedure was discussed in Sect. 3.1.3.

5.3 Information Interaction in Electric Grid

In electric grids, there are two main objects for information interaction: substations (SS) and dispatch centers (DC). Directions for transmitting dispatching information: SS-DC; for administrative and technological information: SS-DS and SS-SS. DC performs the functions of coordinating actions of substation staff in case of emergency situations; DC hosts data acquisition servers of the telecontrol system and power metering system. In the 35 and 110 kV overhead line segment, DC is often located at a large tie substation, or the tie substation is connected to a separate DC by means of wideband communication channels.

As a rule, district electric grid has a treelike topology – power transmission lines diverge from the tie substation and form a sequential chain of substations. There may be taps on the overhead line at the ends of which there are either branch substations or other electric power facilities or consumers.

For dispatcher voice services, the DC private automatic branch exchange (PABX) assigns own telephone number for each substation. The type of connection between DC and SS is force or dial two-wire or four-wire telephone connections of the highest priority using E&M or CAS signaling in E1.

When using IP telephony at substations and DC, IP phones which work with IP PABX in DC are installed, or connection of IP PBX DC-SS/SS-SS via IP network is implemented. As for organization of IP telephony, there are many options for solving this problem (the main – is digital network availability). Very promising for departmental networks of electric power companies is the use of IP phones which can work in peer to peer mode. Phone calls can be made between individual phones using internal routing tables. In this case, IP PBX failure allows saving dispatcher control. At a minimum, these phones can be used for technological communication.

Transmission of telecontrol data should be carried out to the dispatcher center from all substations simultaneously. Data transmission using IEC 60870-5-101 protocol requires arrangement of point-to-point or point-to-multipoint connections between DC and RTU at substations via RS-232/RS-485 interfaces. Data transmission using IEC 60870-5-104 protocol allows data to be transmitted from multiple substations via single physical Ethernet connection using IP routing. All said above also applies to the data transmission to the power metering system.

Information signals are divided into critical and noncritical to the delay in the information delivery:

1. Real-time signals including dispatching and technological communications. The more transmission delay, the worse speech perception quality. It will be shown in the following sections that speech transmission delay is the main limiting factor in creating transit DPLC channels. Typical delay of speech transmission in DPLC channels should not exceed 150 ms for dispatcher voice channels and 250 ms for technological voice channels. Nevertheless, technological telephone DPLC channels in which speech transmission delay exceeds 400 ms exist and are successfully operated in practice.
2. Data signals which are critical to delivery time. These signals include data from the telecontrol system. Delay in transmitting TM and TI signals should not exceed few seconds as a rule.
3. Data signals which are not critical to delivery time. These signals include data of power metering system is transmitted 2–4 times per hour and also data from auxiliary services.

Let us show how the required bit rate is calculated for transmitting speech and data signals.

As already mentioned, low-rate G.723.1 and G.729 speech codecs which form speech frames with a given frequency are mainly used for transmitting speech signals. For G.723.1, the frame length is 30 ms; for G.729, it is 10 ms. When MP-MLQ compression is used, G.723.1 speech frame size is 189 bits, and for ACELP compression – 159 bits. For G.729 codec and CS-ACELP compression, the speech frame size is 80 bits; for G.729D codec and CS-ACELP LPC compression – 64 bits. By multiplying the number of speech frames generated by the speech codec in a second by the frame size, the following required bit rate values can be obtained: G.723.1 MP-MLQ, 6.3 kbps; G.723.1 ACELP, 5.3 kbps; G.729 CS-ACELP, 8.0 kbps; and G.729D CS-ACELP LPC, 6.4 kbps.

If speech is transmitted via internal multiplexers and voice modules of DPLC equipment, the above values should be multiplied by 1.1. Then we get:

G.723.1 MP-MLQ, 7.0 kbps; G.723.1 ACELP, 5.9 kbps; G.729 CS-ACELP, and 8.8 kbps и G.729D CS-ACELP LPC, 7.2 kbps.

But if speech is transmitted using VoIP technology, the voice frame rate of the speech codec should be summarized with the rate of VoIP packets header transmission.

The formula for calculating the required transmission bit rate B_{VoIP} in this case has a form:

$$B_{\text{VoIP}} = 1.05\frac{8\left(S_{\text{header}} + qS_{\text{payload}}\right)}{qt_{\text{frame}}}$$

where q is the number of voice frames in a single VoIP packet; S_{header} is total size of the L2, L3, and L4 headers or compressed header, bytes; S_{frame} – size of the speech frame, and bytes; t_{frame} – duration of one speech frame of the codec, s.

Coefficient 1.05 corresponds to increase of VoIP voice signal transmission bit rate due to the need for parallel transmission of information from the session control protocol – RTCP (real-time control protocol). As a rule, when transmitting VoIP speech, 1 or 2 voice frames of G.723.1 speech codec, and 2 or 3 voice frames of G.729 speech codec are transmitted in a single packet. Figure 5.1 shows structure of the VoIP packet and Ethernet frame.

Using the data shown in Table 3.2, let us calculate G.729 speech codec bit rate for the case of three speech frames in one VoIP packet.

The size of Ethernet and IP packet headers without compression (see Fig. 3.7) will be: 26 bytes, Ethernet; 20 bytes, IPv4; 8 bytes, UDP; 12 bytes, and RTP. Payload size is 30 bytes; total duration of three speech frames in a VoIP packet is $3 \cdot 10 = 30$ ms. Then B_{VoIP} bit rate is equal to $1.05 \cdot 8(66 + 30)/0.03 = 26.9$ kbps, resulting value is more than three times more than rate required for speech transmission via voice modules of DPLC systems. Let us calculate the same for serial WAN connection (see Fig. 3.11); the header sizes remain the same, but instead of Ethernet, PPP protocol is used with a header size of 8 bytes. In this case, B_{VoIP} will be equal to $1.05 \cdot 8(48 + 30)/0.03 = 21.9$ kbps.

To reduce B_{VoIP} value, header compression is applied. For example, after compressing ROHC + EthHC, the header size will decrease to 18 bytes, and B_{VoIP} will be 13.5 kbps. Similarly, for ECRTP and PPP compression at a total header length of 20 bytes, B_{VoIP} will be equal to 14.0 kbps.

The rate of the data transfer through point-to-point connections and RS-232 interface when, for example, transmitting data from telecontrol system using the IEC 60870-5-101 protocol, is the same for each connection. The values 2.4, 4.8, and 9.6 kbps are often used. The data frame consists of 1 start bit, 8 payload bits, 1 parity check bit, and 1 stop bit. Thus, 3 bits of 11- bit frame account for service information. It is easy to calculate how much useful data can be transmitted per second; at a connection rate, for example, of 9.6 kbps, it will be $9600\ 8/11 = 6980$ bits or 872 bytes. In IEC 60870-5-101 standard, FT1.2 protocol is used to control data transmission. FT1.2 protocol frame includes 8 bytes in service information fields and up to 255 bytes in APDU (application protocol data unit) field with maximum value of ASDU (application service data unit) useful information field of 253 bytes, of which 4 bytes are the address of the information object. We get by simple calculations that,

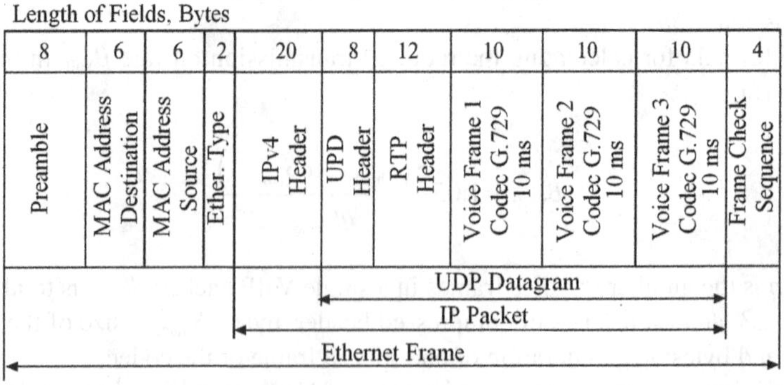

Length of Fields, Bytes

8	6	6	2	20	8	12	10	10	10	4
Preamble	MAC Address Destination	MAC Address Source	Ether. Type	IPv4 Header	UPD Header	RTP Header	Voice Frame 1 Codec G.729 10 ms	Voice Frame 2 Codec G.729 10 ms	Voice Frame 3 Codec G.729 10 ms	Frame Check Sequence

UDP Datagram

IP Packet

Ethernet Frame

Fig. 5.1 Ethernet frame with VoIP packet

in 1 second at specified parameters, telemetering signals can be transmitted with a time stamp 872·249/(263·15) = 55; telemetering signals without a time stamp 872·249/(263·8) = 103. It should be noted that in practice, when transmitting data of telecontrol via DPLC channels, a smaller value of the ASDU data field is used – 100–120 bytes, and TI and TM signals are often transmitted without time stamps.

When transmitting data of telecontrol via Ethernet networks at the transport layer, IEC 60870-5-104 standard provides for use of TCP guaranteed delivery protocol. IEC 60870-5-104 protocol has its own algorithms for verifying data delivery. The standard provides selecting number of data blocks (APDU) k which the sender can send before acknowledgment of their reception.

Checking is performed at the application level, and parameter k does not affect operation of the TCP transport layer protocol. Receiver sends acknowledgments at least after receiving w packets. Parameters k and w can have a value from 1 to 32767. It is recommended that w value does not exceed two-thirds of k value. By default, k value is 12. Figure 5.2 shows information frame structure for transmitting data using IEC 60870-5-104 protocol.

Information frame includes headers of Ethernet, IP, and TCP protocols. Maximum size of MTU (maximum transmission unit) for Ethernet protocol is 1500 bytes, and minimum size is 46 bytes. MTU transmits the IP packet itself, with 20 byte IPv4 header which in turn transmits TCP segment with 20 bytes headers and payload. Size of the TCP segment is defined by such parameter as maximum segment size (MSS), which is responsible for the size of payload. If MSS size is more than the size prided for the second-layer MTU protocol, the IP packet will be fragmented during data transfer. This should be avoided in order to make better use of the channel's bandwidth. In most cases, in computer networks where the data bit rate is calculated in megabits per second, MTU size of the Ethernet frame is 1500 bytes, and MSS size is 1460 bytes; for wireless connections, MTU size is 576 bytes, and MSS size is 536 bytes, respectively. For example, when configuring ASDU with a size of 200 bytes (APDU 202 bytes) and MSS size of 1460 bytes, 7 APDU data blocks are transmitted in a single IP packet (provided that $k > 7$).

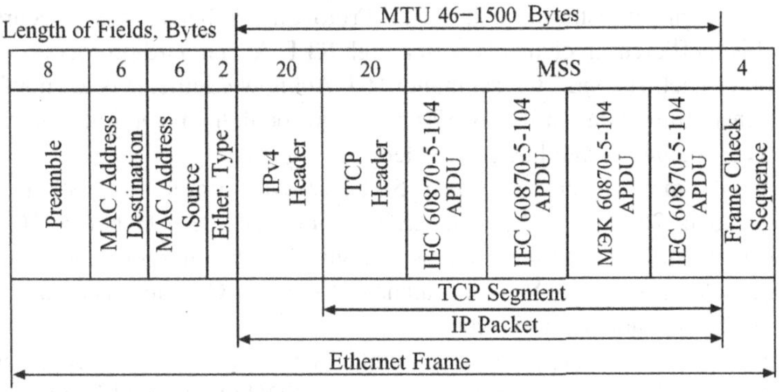

Length of Fields, Bytes

MTU 46–1500 Bytes

8	6	6	2	20	20	MSS				4
Preamble	MAC Address Destination	MAC Address Source	Ether. Type	IPv4 Header	TCP Header	IEC 60870-5-104 APDU	IEC 60870-5-104 APDU	МЭК 60870-5-104 APDU	IEC 60870-5-104 APDU	Frame Check Sequence

TCP Segment

IP Packet

Ethernet Frame

Fig. 5.2 Ethernet frame with TCP packet

As a rule, RTU of telecontrol systems provides MSS size adjustment. Accordingly, by changing ASDU and MSS values, it is possible to achieve the best data transfer via the communication channel. Adjustment of the size of transmitted packets can be also carried out by configuring the small size of TCP receive window at the receiving computer.

The use of large MSS values in low-speed channels with bit rates of tens of kilobits per second leads to the following problems. First, the packet serialization time when it is transmitted via linear interface of the DPLC system becomes large. For example, if the Ethernet frame length with a header is equal to 1526 bytes and bit rate of a DPLC equipment modem is 32 kbps, the serialization time is $8 \cdot 1526/32 = 380$ milliseconds. If transmission of VoIP speech in parallel with the data is planned, this delay in waiting for the VoIP packet in the transmission queue is unacceptably large. Second, the larger the data packet, the more probable this packet damaging by errors. On the other hand, high MSS value ensures efficient data transfer: in this case, a portion of overhead traffic does not exceed 5%. Reducing the MSS size reduces size of information frame and, consequently, reduces probability of the frame damaging by errors. But this increases portion of overhead traffic and requires compression of IP packet headers.

In [3], selecting optimal MTU value for data transmission via channel with errors for the case of data packets with full-size headers using IPCH compression or using robust ROHC-TCP compression was considered. The problem of minimizing the actual volume of transmitted data V_{sum} was solved under conditions that the total volume of transmitted data is 10^6 bytes, and optimization parameter is the payload length $S_{payload}$ (Fig. 5.3); here, MFP is framed transmission when the session context is updated with specified interval F packets.

The graph shows that optimal MTU size for transmitting data via DPLC channels without header compression is 150–300 bytes, and for MSS – 110–260 bytes. When using robust header compression, recommended MTU size is 120–200 bytes, and MSS size at summarized length of compressed IP and TCP headers of 7 bytes is 113–193 bytes.

Let us recall that the lower the MSS value, the lower delay in transmitting speech packets due to the waiting time in the service queue.

Since data are transmitted using TCP Protocol, another problem occurs. TCP protocol is inefficient in channels with a high RTT. At standard protocol operation, reception of each second TCP segment (TCP Acknowledgment) is acknowledged. When transmitting data via IEC 60870-5-104 protocol, the reception of each transmitted TCP segment should be acknowledged (Fig. 5.4).

Let us assume that the IEC 60870-5-104 Ethernet frame size without header compression is 266 bytes. At that, ASDU value will be 198 bytes. TCP ACK acknowledgment will be sent in packets without a load when information frame size is 66 bytes (headers only). Value of latency of the DPLC system is assumed to be 60 ms; bit rate of modems is 32 kbps.

With such initial data, round trip time is approximately $2 \cdot 60 + 8(266 + 66)/32 = 200$ ms. That is, only 5 APDU blocks or 990 bytes of information (ASDU) with TM and TI data can be transmitted in 1 second. It means, that

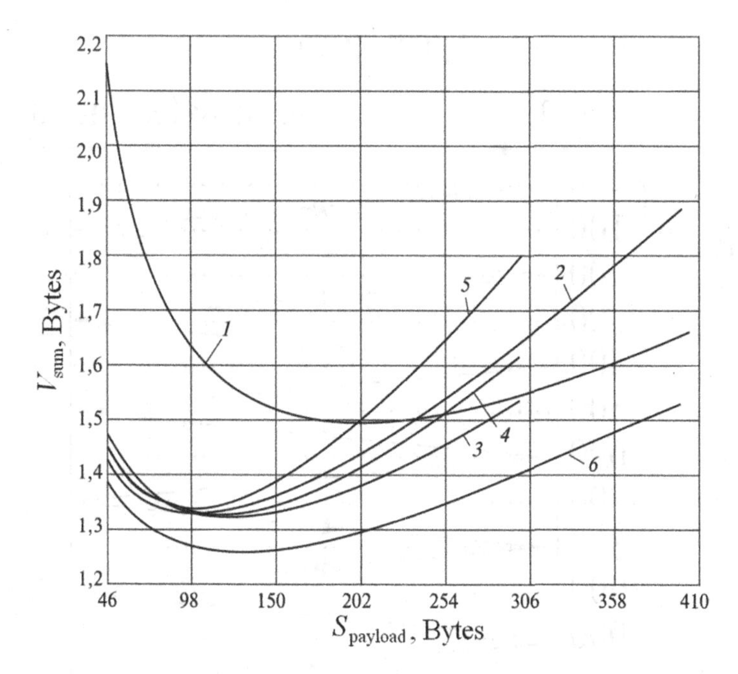

Fig. 5.3 Graphs of $V_{sum}(S_{payload})$: 1, without header compression; 2, IPHC without framing; 3-IPHC with MFP, $F = 6$; 4-IPHC with MFP, $F = 10$; 5-IPHC with MFP, $F = 20$; 6-robust header compression. (© 2020 IEEE. Reprinted, with permission, from Merkulov and Shuvalov [3])

channels throughput are only 8·990/1000 = 7.92 kbps, while DPLC modems bit rate is 32 kbps. The inability of TCP protocol to use a group acknowledgment of several segments reception is the main problem at telecontrol data transmission via IEC 60870-5-104 protocol. This function should be implemented directly in the RTU. To improve channel throughput for IP DPLC channels, it would be helpful to use approaches described in [4].

We encounter similar problems when transmitting data from auxiliary services. When data are transferred between computers, MTU and MSS can be reduced by configuring required MTU size in the operating system registry. Then IP packets in which MSS exceeds MTU value will be fragmented. RTT can be reduced by configuring TCP ACK parameters – number of received packets – in the operating system registry that is followed by acknowledgment. As an example, Fig. 5.5 shows data transfer process when sending acknowledgment after receiving eights packets.

Having considered main types and characteristics of information signals in DPLC channels, we will proceed to the discussion of types of DPLC channels and to calculation of required modem bit rate.

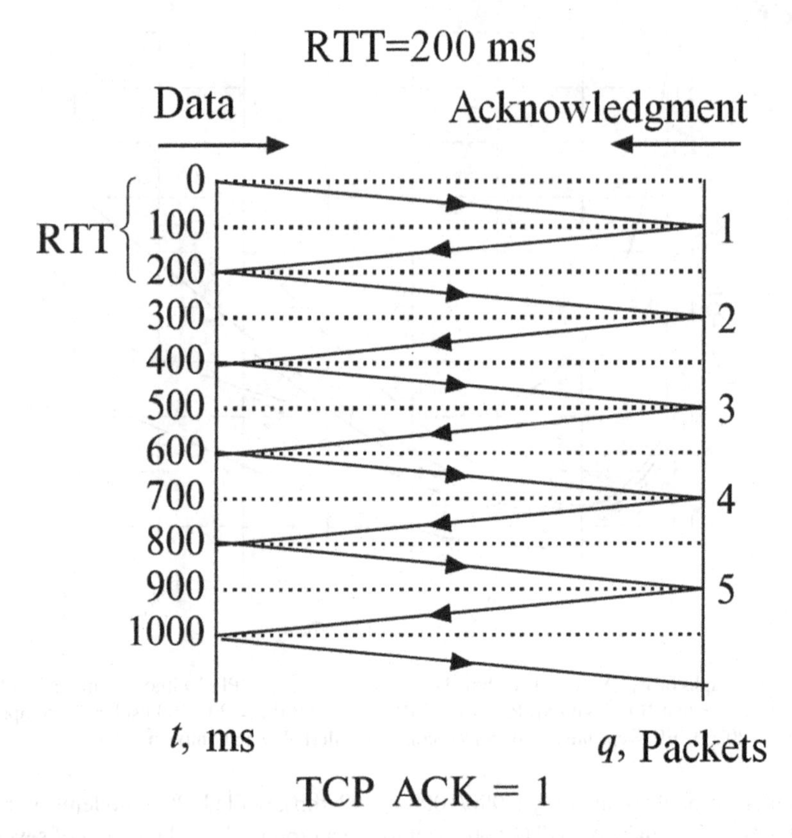

Fig. 5.4 Diagram of TCP data transfer when acknowledging each packet

5.4 Types of DPLC Channels

According to the method of information signal transmission, DPLC channels can be divided into three types:

- DPLC channels with time division multiplexing. Data and speech signals from the DPLC equipment interface modules are processed in a TDM multiplexer, and their transmission delay is constant.
- DPLC channels with time division multiplexing of high-priority signals and packet transmission of data signals. Delay in transmitting high-priority signals is constant. Packet queues may occur for data transmitted via Ethernet-bridge or WAN connection, and in this case, delivery time is variable.
- DPLC channels with packet transmission of speech and data traffic. In comparison with the second variant, the task of transmitting high-priority information is more complicated due to the need to ensure low delay in its transmission.

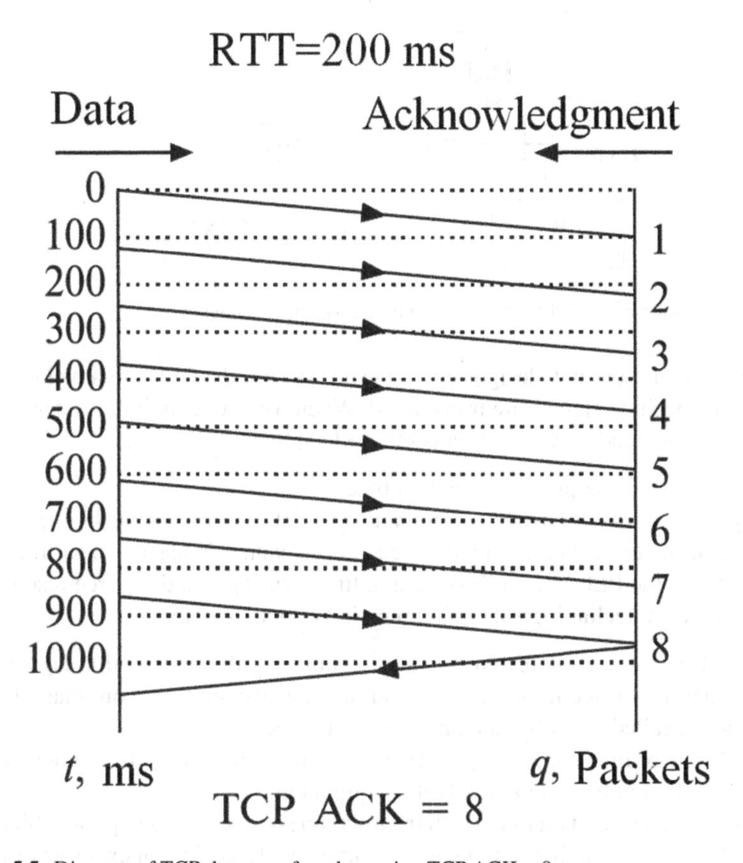

Fig. 5.5 Diagram of TCP data transfer when using TCP ACK = 8

All three types of DPLC channels can be used in the telecommunication network of electric power grid company and, accordingly, for each of them should be used own methods for calculating required bit rate and delay at signal transmission.

5.4.1 *DPLC Channel with Time Division Multiplexing of Information Signals*

DPLC channel with time division multiplexing (Fig. 5.6) is the simplest in terms of calculating required bandwidth and transmission delay. Speech signals are transmitted using a built-in multiplexer and voice modules. Data signals are transmitted via asynchronous RS-232 connections, for example, data from telecontrol system using the IEC 60870-5-101 protocol.

Information signal prioritization is performed by the multiplexer itself depending on specific design of the DPLC equipment. For example, for each information signal or interface, specific number corresponding to its priority is assigned.

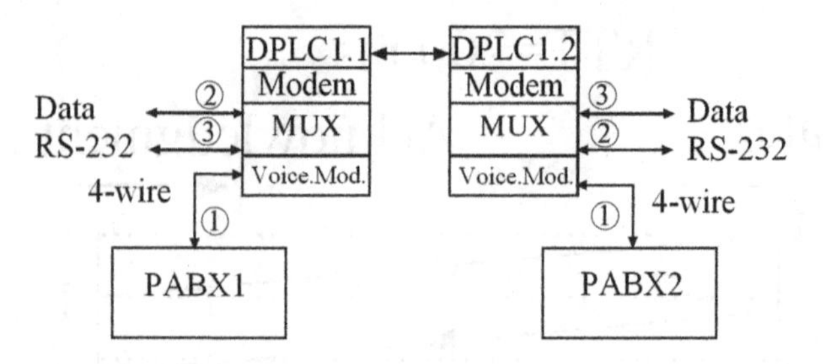

Fig. 5.6 Block diagram of DPLC channel with time division multiplexing

Under normal (nominal, design) operating conditions, the DPLC channel is fully loaded – all specified signals are transmitted. When weather conditions deteriorate, the bit rate of the modem connection is reduced. At that:

- Transmission of low-priority signals is blocked.
- Low-priority signal transmission remains possible if there is no high-priority signal transmission. For example, since voice channel is usually busy for less than 30% of the PLC channel's operating time. During idle time, free resources can be used for technological communication or data transfer.

Required bit rate is calculated by simply summing the information signal bit rates considering service information of multiplexor (usually not more than 10%). This bit rate is called gross bit rate on the HF interface.

Total bit rate of information signals transmission cannot exceed the values of bit rate available for configuring in the DPLC equipment.

For example, required bit rate for all information signals is 27.5 kbps, and bit rate available for configuring in the modem is 28 kbps. In Table 5.3, an example of this DPLC system configuration, a system with three steps of adaptation, is given.

In the most advanced DPLC systems, technological voice channel can work at 2-nd 3-rd steps of adaptation when dispatcher voice channel is idle, and power metering data can transmit at 3-rd step when voice channels are idle.

When using DPLC channels with time division multiplexing, delay in transmitting information signals will always be constant. For data signals, it is equal to the latency of DPLC systems $t_{Dx} = t_{DPLC}$. For speech signals, t_{Vx} consists of latency of DPLC systems t_{DPLC} and speech codec delay t_{codec}:

$$t_{Vx} = t_{DPLC} + t_{codec} = t_{DPLC} + \left(t_{frame} + t_{alg}\right)$$

where t_{frame} is duration of the speech frame; t_{alg} is speech codecs algorithmic prediction delay.

Latency of DPLC equipment is defined using technical characteristics provided by the manufacturer. The speech frame duration for G.729 codec is 10 ms; algorithmic prediction delay is 5 ms, i.e., the codec delay is 15 ms. The speech frame duration for G.723.1 codec is 30 ms, algorithmic prediction delay is 7.5 ms, respectively, and the codec delay is 37.5 ms.

Table 5.3 Example of bit rate calculation for DPLC channel with time division multiplexing

Signal	Priority	Bit rate, kbps	Modem operation parameters
Dispatcher voice (DispVoice) G.729 CS-ACELP LPC	High	Steps 1–3: 7.2	Normal operating mode – 1 step: 27.5 kbps; $B_{DPLC1} = B_{DispVoice} + B_{TechVoice} + B_{TC} + B_{PM}$
Telecontrol (TC) (RS-232) IEC 60870-5-101	High	Steps 1–3: 9.6	
Power metering (PM) (RS-232)	Medium	Steps 1–2: 4.8 Steps 3: switched off	Adaptation – 2 step: 21.6 kbps; $B_{DPLC2} = B_{DispVoice} + B_{TC} + B_{PM}$
Technological voice (TechVoice) G.723.1 ACELP	Normal	Step 1: 5.9 Steps 2–3: switched off	Adaptation – 3 step: 16.8 kbps; $B_{DPLC3} = B_{DispVoice} + B_{TC}$

If, for example, the DPLC equipment latency $t_{\mathrm{DPLC}} = 60$ ms, speech delay when using G.729 codec is 75 ms or 97.5 ms for G.723.1, respectively. Data transfer delay is 60 ms.

5.4.2 DPLC Channel with Time Division Multiplexing of Information Signals and Packet Traffic Transmission

The second type of DPLC channel includes functions for transmitting packet traffic using Ethernet connection.

Switch or router is an external network device which is connected to the DPLC equipment either via synchronous serial interface X.21 (Fig. 5.7a) or via DPLC equipment network element by means of Ethernet bridge interface (Fig. 5.7b).

Voice signals are processed by built-in voice modules and DPLC equipment multiplexer.

Data transmission via asynchronous RS-232 connections may be completely not used. Data from telecontrol and power metering are transmitted using IEC 60870-5-104 protocol and can be connected either to external network switches or routers or to Ethernet interface of the built-in network element of the DPLC equipment.

Similarly to the first variant of DPLC system structure, delay of data transmission when using RS-232 interface will remain equal to latency of the DPLC equipment. Delay of voice transmission, as before, will consist of latency of the DPLC system and codec delay.

For packet data transmitted via Ethernet connection, another source of delay should be taken into account – packet serialization time when the packet is transmitted via DPLC channel $t_{\mathrm{serial}.D}$. If there are n packages in the service queue and transmission of the first one has already started, second package will be served only after the first package has been serviced. Within the queue, packets can be rearranged according to their priority (see below), but not after transmission beginning.

Information signal prioritization is performed in the same way as in the previous example, but packet traffic prioritization technologies are added to simple disabling/enabling of services in the multiplexer.

Let us return to the example of modem bit rate adaptation shown in Fig. 3.4. Data on information signal and modem bit rate given in Table 5.4 are based on this example. It can be seen that the data bit rate for Ethernet connection varies from 37.6 kbps, for the first adaptation step to 12.0 kbps for the fourth adaptation step. At the 5th adaptation step, Ethernet data will be transmitted only if the dispatcher voice channel is idle. At step 6, Ethernet data transmission is blocked.

Serialization delay $t_{\mathrm{serial}.D}$ is calculated by dividing the information frame full size (with all headers) by Ethernet (service) data bit rate B_{Eth} inside the DPLC equipment multiplexer:

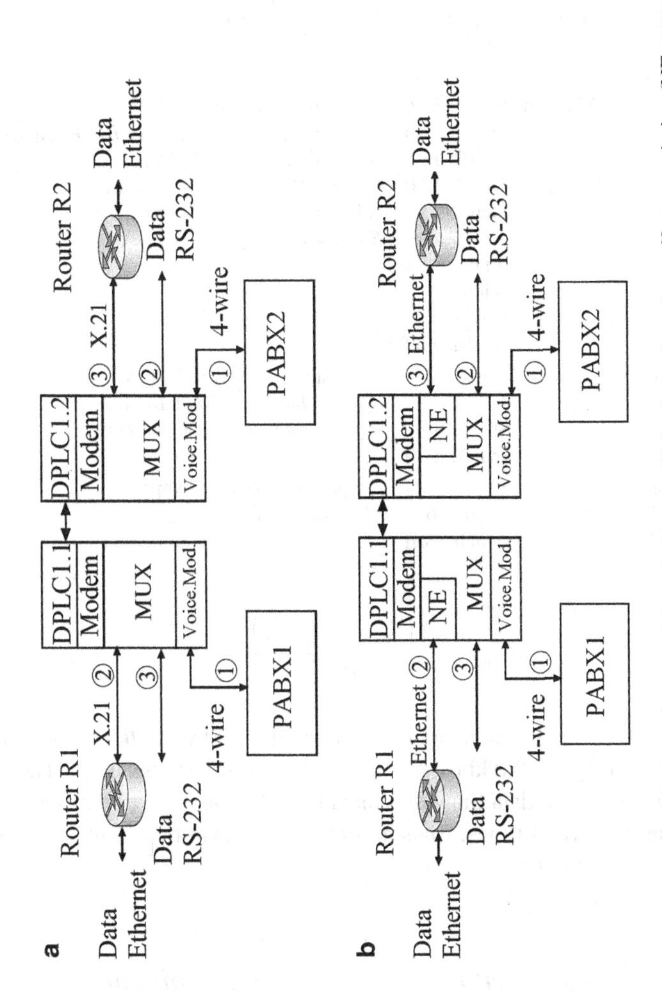

Fig. 5.7 Block diagram of DPLC channel with information signal time division multiplexing and packet traffic transmission (NE-network element)

Table 5.4 Example of calculation of modem bit rates for DPLC channels with time division multiplexing and packet traffic transmission

Signal	Priority	Bit rate, kbps	Modem operation parameters
Dispatcher voice (DispVoice) G.729 CS-ACELP LPC	High	Steps 1–6: 7.2	Normal operating mode – 1 step 64 kbps; $B_{\text{DPLC1}} = B_{\text{DispVoice}} + B_{\text{TechVoice1}} + B_{\text{TechVoice2}} + B_{\text{TC}} + B_{\text{PM,AUX}}$
Telecontrol (TC) (RS-232) IEC 60870-5-101	High	Steps 1–6: 4.8	
Power metering (PM) and other data services (AUX)	Medium	Step 1: 37.6 Step 2: 25.6 Step 3: 18.8 Step 4: 12.0 Step 5: 4.8 + 7.2 if DispVoice is idle Step 6: switched off	Adaptation – 2 step: 52 kbps; $B_{\text{DPLC2}} = B_{\text{DispVoice}} + B_{\text{TechVoice1}} + B_{\text{TC}} + B_{\text{PM,AUX}}$ Adaptation – 3 step: 38 kbps; $B_{\text{DPLC3}} = B_{\text{DispVoice}} + B_{\text{TC}} + B_{\text{PD,AUX}}$ Adaptation – 4 step: 24 kbps; $B_{\text{DPLC4}} = B_{\text{DispVoice}} + B_{\text{TC}} + B_{\text{PM,AUX}}$
Technological voice #1 (Tech.Voice1) G.729 CS-ACELP LPC	Normal	Steps 1–2: 7.2 Steps 3–4: switched off	Adaptation – 5 step:16.8 kbps; $B_{\text{DPLC5}} = B_{\text{DispVoice}} + B_{\text{TC}} + B_{\text{TC,AUX}}$; $B_{\text{PM,AUX}}$—at resources availability
Dispatcher voice #2 (DispVoice1) G.729 CS-ACELP LPC	Low	Step 1: 7.2 Steps 2–6: switched off	Adaptation – 6 step:12 kbps; $B6 = B_{\text{DispVoice}} + B_{\text{TC}}$

$$t_{\text{serial}.D} = \frac{8\left(S_{\text{header}} + S_{\text{payload}}\right)}{B_{\text{Eth}}}$$

For example, for Ethernet frame with a length of 266 bytes, B_{Eth} = 37.6 kbps $t_{\text{serial}.D} \approx 57$ ms, and $B_{\text{Eth}} = 12.0$ kbps, $t_{\text{serial}.D} \approx 177$ ms – three times more. The main advantage of using packet data transmission is that, despite reduction of the DPLC modem bit rate by several times, transmission of data continues, but surely with increased transmission delay.

5.4.3 *DPLC Channel with Packet Traffic Transmission*

A rather interesting issue is creation of converged DPLC channels, where, both speech and data are transmitted in packets.

It should be noted that such networks were previously created using Frame Relay and ATM technologies and still function successfully [5–8]. In Frame Relay DPLC networks, FRAD (Frame Relay Access Device) multiplexers with FXS/FXO/E&M voice modules and RS-232 low-speed modules were used as switching equipment.

Table 5.5 Classes of packet traffic in DPLC channels

Type of traffic	PHB encoding	CoS	Examples of application:
Network control	CS6, CS7	6,7	Routing protocols EIGRP, OSPF, etc.
VoIP	EF	5	Dispatcher voice. Technological voice
Multimedia	AF41/42/43	4	VoIP signaling protocol
Technological data	AF31/32/33	3	Telecontrol, power metering
Other data	AF21/22/23	2	E-mail, file exchange
Other services	CS0 /BE	0	Internet

PHB – per-hop forwarding behavior (code which shows how to process a packet in a transit router at the path from the source to the recipient). CS (class selector) – selection of service class. EF (expedited forwarding) – priority delivery with minimal delay, jitter, packet loss, and guaranteed transmission resources. AF (assured forwarding) guaranteed delivery, the higher the number, the higher the priority of the package. CS0 /BE (best effort) – non-guaranteed delivery. CoS (class of service) – service class

ATM switches were used as network equipment in ATM DPLC networks. In both cases, network equipment was connected to the DPLC systems via X.21 synchronous serial interface. Speech was transmitted using VoFR (voice of frame relay) and VoATM (voice over ATM) technologies. Currently, lifecycle of such networks has ended due to the end of lifecycle of Frame Relay and ATM technologies, and their place in packet-switched network creation has been taken by IP technologies.

Network equipment is connected to the DPLC systems either via built-in network elements when Ethernet bridge is used (Fig. 5.8a) or via X.21 synchronous serial interface when using synchronous WAN connection (Fig. 5.8b). In both cases, entire available bit rate of the DPLC channel is used for transmitting packet traffic, regardless of whether speech or data packets are being transmitted.

Signal prioritization is the most difficult task in such channels. Packets of various services, including VoIP, enter the switch from different Ethernet ports and are transmitted to the DPLC equipment via a single port. Ethernet interface bit rate is 10 or 100 Mbps, and bit rate of DPLC equipment is tens or hundreds of kilobits per second. Therefore, delay in packet serialization by Ethernet interfaces makes microseconds and in DPLC equipment – tens of milliseconds.

Packet traffic prioritization can be performed inside DPLC equipment, as described in Sect. 3.2.2. But it should be said that the main role in the prioritization process should be played by external devices – routers, and for DPLC channel with synchronous serial WAN connections, only external router can be responsible for prioritizing packets in general.

To prioritize packets in IP networks, a special identifier is used, DCSP (Differentiated Service Code Point), located in the TOS (Type of Service) field of the IP protocol header. Entire DCSP traffic is divided into seven classes by priority and service parameters [9]. Information about packet traffic classes which can be used for PLC channels is given in Table. 5.5.

The bit rate of DPLC modems affects delay of information transmission: the lower the modem bit rate, the more serial delay in the queue of information frames, especially during periods of dynamic adaptation.

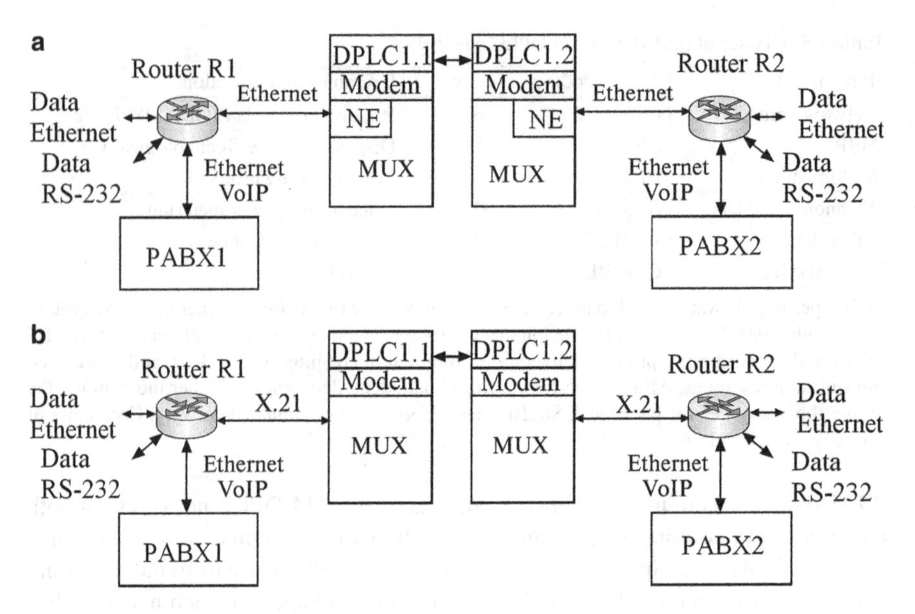

Fig. 5.8 Block diagram of a DPLC channel with packet speech and data traffic transmission (NE-network element)

Let us consider a case when two VoIP signals and one data signal are transmitted via DPLC equipment. Speech processing is performed in a VoIP gateway built into the router, G.729 codec is used, and three speech frames are transmitted in a single VoIP packet. Data is transmitted in Ethernet frames of 266 bytes with each TCP segment acknowledgment, and transmission frequency is at least five data packets per second. Let us assume that the DPLC equipment is connected to the router via X.21 synchronous serial interface (see Fig. 5.8b) with PPP encapsulation of Ethernet frames and compression of VoIP – ECRTP and TCP – IPHC packet headers (see Table 3.2). Latency of the DPLC equipment is 60 ms. It is required to find minimum bit rate of the DPLC channel enough for simultaneous transmission of speech and data and delay in transmitting speech packets.

Calculation of the speech signal bit rate was carried out in Sect. 5.2, and for the case in question, for transmitting one or two speech signals, the bit rate will be $B_V = 14$ or 28 kbps, respectively. This bit rate is the minimum required for speech transmission, provided that transmission of data packets is completely blocked during speaking, since VoIP packets have high CoS = 5 and class of service EF, and data packets have CoS = 3 and class of service AF.

As for the data transmission, if it is necessary to transmit five data packets per second and RTT time should not exceed 200 ms, its value cannot be less than $B_D = 5{\cdot}8{\cdot}215/1000 = 8.6$ kbps total bit rate of the connection will be $B_{DPLC} = 28 + 8.6 = 36.6$ kbps.

Let us evaluate whether B_{DPLC} value was found correctly and whether condition RTT < 200 ms is met.

As the packets may stay in the service queue on their way from the source to the destination, their delay will increase.

For example, if VoIP packets of two voice connections and data session packets are transmitted simultaneously in the DPLC channel, the voice packet with the highest priority will still be queued:

- Behind a data packet transmitting of which has already started
- Behind a previously received voice packages, transmission of which has already started
- Behind a voice packet in a parallel session waiting for the data packet to be transmitted

In turn, data packet with a priority lower than the voice packet always stay in the service queue behind one or two voice packets.

Deviation of arrival time of the packets or jitter in the general case is within:

- For voice packets:

$$0 \le t_{\text{jitter}} \le= (n-1)t_{\text{serial}V} + t_{\text{serial}D}$$

- For data packets:

$$0 \le t_{\text{jitter}} \le nt_{\text{serial}V}$$

At that:

$$\text{RTT} = 2t_{\text{DPLC}} + t_{\text{serial}D} + t_{\text{serial}ACK} + nt_{\text{serial}V}$$

where $t_{\text{serial.}ACK}$ is the packet serialization time with acknowledgment TCP – ACK that the packet has been transmitted from the receiver to the source.

Let us calculate RTT for specified parameters:

2·60 + 8·215/36.6 + 8·15/36.6 + 2·8·50/36.6 ≈ 192 ms. This means that B_{DPLC} value is selected correctly and can even be slightly less-33 kbps.

Delay in end-to-end speech transmission:

$$t_{\text{EED}} = t_{\text{DPLC}} + t_{\text{serial}V} + t_{\text{codec}} + t_{\text{jitter}}$$

It should be considered at calculation that, in VoIP technology, speech codec delay should be found from the formula [10]:

$$t_{\text{codec(VoIP)}} = (q+1)t_{\text{frame}} + t_{\text{alg}}$$

In this case, $t_{\text{codec(VoIP)}} = (3 + 1)\cdot 10 + 5 = 45$ ms, where $q = 3$ is number of speech frames transmitted in a single packet.

When designing networks with packet switching, maximum possible delay t_{jitter} is used; in this case, $t_{\text{serial}.V} = 8\cdot(20 + 30)/36.6 \approx 11$ ms, $t_{\text{serial}.D} = 8\cdot215/36.6 \approx 47$ ms and, respectively, $t_{\text{jitter}} \approx 58$ ms.

As a result, maximum delay in end-to-end speech transmission is:

$$t_{\text{EED}} = 60 + 11 + 45 + 58 = 174\text{ms}.$$

Let us estimate the number of data packets which could be transmitted if the voice channels are not busy: RTT = $2\cdot60 + 8\cdot215/36.6 + 8\cdot15/36.6 \approx 170$ ms, i.e., q_{D} value would not increase even by one packet.

At the same time, if the channel was used only for speech transmission, t_{EED} value would be significantly reduced to 116 ms.

It can be said based on the above calculations that in converged IP-DPLC channels, increase of the DPLC modem bit rate B_{DPLC} to the values more than required one does not lead to the expected improvement in data transmission parameters and throughput, since main sources of delay in the channel are DPLC equipment and speech codecs.

When two packets of VoIP sessions are transmitted in parallel and VAD (voice activity detection) algorithm is used to suppress speech pauses, the data packet does not get in the queue behind two voice packets; it manages to "jump" during pauses in conversations. VAD function allows reducing required total bit rate when number of connections is more than 20 that is only possible in digital PLC channels with a bit rate of hundreds of kbps. Since this number of simultaneous connections is not typical for DPLC channels, VAD should not be considered account when calculating required gross bit rate for the HF interface. VAD function is useful because data packets can be transmitted during speech pauses.

In principle, when saying about DPLC channels use in IP networks, it should be recognized that measures to improve the DPLC system technical characteristics are necessary. Primarily, in the following areas:

- Reducing DPLC equipment latency up to 15... 20 ms.
- Implementing of robust header compression on data link, network, and transport layers.

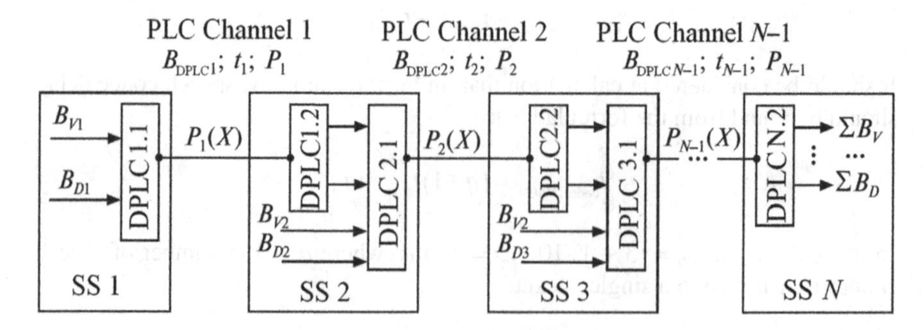

Fig. 5.9 The simplest model of DPLC channel with accumulating load

- Implementing of payload compression to improve channel throughput.
- Investigation of separate speech and data packet transmission, for example, using DMT systems, when several carrier frequency groups are used for speech transmission, and the remaining groups – for data transmission. If there are no active VoIP sessions, all frequency groups are used for data transmission.
- Implementing algorithms for division of large frames into smaller fragments by means of internal functions of the DPLC systems. Development of these areas of DPLC communication technology is the subject of a special study which is promising from the point of view of mass transition to DPLC channels application for IP networks.

5.5 DPLC Channels with Transit Stations

Planning DPLC channels with several transit stations is a rather common design task. Let us consider issues of planning such channels and show what limitations on the transit channels number exist.

5.5.1 The Simplest Model of DPLC Channel with Transit Stations

Figure 5.9 shows the simplest scheme of transit DPLC channel with a consistently increasing load in which information from remote substations is transmitted to the dispatch center via a single channel between each of the neighbor substations of the power system. A special case of this scheme is transfer of information to the dispatch center from a single remote substation and through several intermediate substations.

Model of such a DPLC channel is described by a system of three inequalities:

$$\begin{cases} \sum_{i=1}^{N-1} B_{\mathrm{V}(i)} + \sum_{i=1}^{N-1} B_{\mathrm{D}(i)} \le B_{\mathrm{DPLC}i} \\ \sum_{i=1}^{N-1} t_i \le t_{\max} \\ \prod_{i=1}^{N-1} P(X)_i > P(X)_{\min} \end{cases}$$

where N is the number of substations in the channel.

Each section of the channel is characterized by the following parameters:

1. Bit rate of modems B_{DPLC} should be more than the sum of the bit rates of all information signals which can be transmitted simultaneously the next station. It

is obvious that the closer the substations and channels connecting them to DPLC to the dispatch center, the more number of information signals – own and transited – transmitted via the DPLC channels.

2. Delay in transmitting information signals *t*. Delay sources for different types of DPLC channels were discussed in the previous section. The more distance from the substation to the dispatch center and more transit substations, the more delay in transmitting information t_{max} between this substation and the dispatch center. As will be shown in the next section, maximum allowed transmission delay is the main parameter which limits the number of transit sections in DPLC channels of the treelike structure.
3. Reliability of *P*(*X*) channel. Using this parameter, unavailability time of the channel can be calculated. Main factors affecting the PLC channel reliability are design, in particular, selection of the PLC channel operating frequencies (see Sect. 5.6), operating conditions of the channel, mean time between failures of PLC equipment hardware, line traps, coupling capacitors and coupling devices, duration of repairs and maintenance, etc.

5.5.2 Limiting Transit Sections Based on the Delay and Speech Transmission Quality Criteria: Characteristics of Speech Codecs

In public switched telephone networks, average expert evaluation of speech quality MOS (mean opinion score) is used to assess speech quality and objective criteria are delay in speech signal transmitting, type of codec, and percentage of speech frame lost.

For DPLC channels, a simplified approach can be applied, consisting of two simple requirements:

End-to-end delay in speech transmission should not exceed the value t_{max}. Practical study shows that, for dispatcher voice channels, t_{max} should not exceed 250 ms, and recommended value is less than 150 ms; for technological voice channels, transmission delay of 250 ms... 300 ms when conducting conversations between substations is not critical for the service staff (voice quality is comparable to cellular communication when conducting conversations between two cities);

Number of speech compression/decompression cycles of speech codecs should not exceed 3. Reason is that speech compression and decompression operations are always made with losses, and decrease of speech quality when such operations number is more than three becomes inaccessible.

Speech codecs based on MP-MLQ algorithm provide a lower level of distortion compared to speech codecs based on CELP algorithm – this is important in the cases where voice signal is switched through a central PBX. G.723.1 codec (MP-MLQ) suites better for creating transit connections; it adds a delay of 37.5 ms

(1 speech frame of 30 ms). G.729 codec (CS-ACELP) has a lower delay of 15 ms (1 speech frame of 10 ms) but worse performance when creating voice frequency interconnections.

Figure 5.10 shows two variants for planning transit DPLC channels which are most often used in practice.

In this case, both limitations apply. The number of transit sections should not exceed the allowed number of compression/ decompression operations 3 – 1 = 2 – one operation is performed at the beginning and end of the channel. End-to-end speech transmission time delay should not exceed $t_{\text{EED}} < t_{\max}$:

$$t_{\text{EED}} = \sum_{i=1}^{N-1} t_{\text{DPLC}i} + (N-1)\,t_{\text{codec}}$$

Variant 1, Fig. 5.10a The simplest variant is when the voice signal is transmitted via the DPLC channels with information signal time division multiplexing, and speech transit connections are carried out via analog voice (VF) interfaces.

For example, let DPLC latency $t_{\text{DPLC}} = 60$ ms, codec delay $t_{\text{codec}} = 15$ ms (G.729 with one speech frame), then for a single transit connection $t_{\text{EED(1)}} = (3 - 1)(60 + 15) = 150$ ms, with two transit connections – $t_{\text{EED(2)}} = (4 - 1)(60 + 15) = 225$ ms. For G.723.1 codec with a delay of 37.5 ms, we get $t_{\text{EED(1)}} = (3 - 1)(60 + 37.5) = 195$ ms and $t_{\text{EED(2)}} = 292.5$ ms. This means that, from the point of view of maximum delay criterion, for dispatcher voice with G.729 codec, in this case, 1 retransmission can be used, for technological voice – two retransmissions with G.729, one – with G.723.1 and two retransmissions with G.723.1, in case, when there is no another possibility to create voice communication.

Variant 2, Fig. 5.10b Voice signal is transmitted via several transit sections, but transit is performed not through the voice frequency interfaces but through DPLC equipment special transit interface at digital mode, which requires performing speech compression and decompression operations only at the ends of the channel. Principle of this scheme operation was discussed in Sect. 3.1.4.

End-to-end delay of speech transmission for this case is as follows:

$$t_{\text{EED}} = \sum_{i=1}^{N-1} t_{\text{DPLC}i} + t_{\text{codec}}$$

Here, for $t_{\text{DPLC}} = 60$ ms, $t_{\text{ocdec}} = 15$ ms and one transit connection, we get $t_{\text{EED(1)}} = (3 - 1)\,60 + 15 = 135$ ms, for two transit connections – $t_{\text{EED(2)}} = (4 - 1)\cdot 60 + 15 = 195$ ms, for three transit connections – $t_{\text{EED(3)}} = (5 - 1)\cdot 60 + 15 = 255$ ms.

When using G.723.1 codec with a delay of 37.5 ms, we get $t_{\text{EED(1)}} = (3 - 1)\cdot 60 + 37.5 = 157.5$ ms and $t_{\text{EED(2)}} = = 217.5$ ms. This means that, from the point of view of maximum delay criterion, up to two transit connections with G.729 codec and one with G.723.1 codec can be used for dispatch voice, while two G.723.1 and

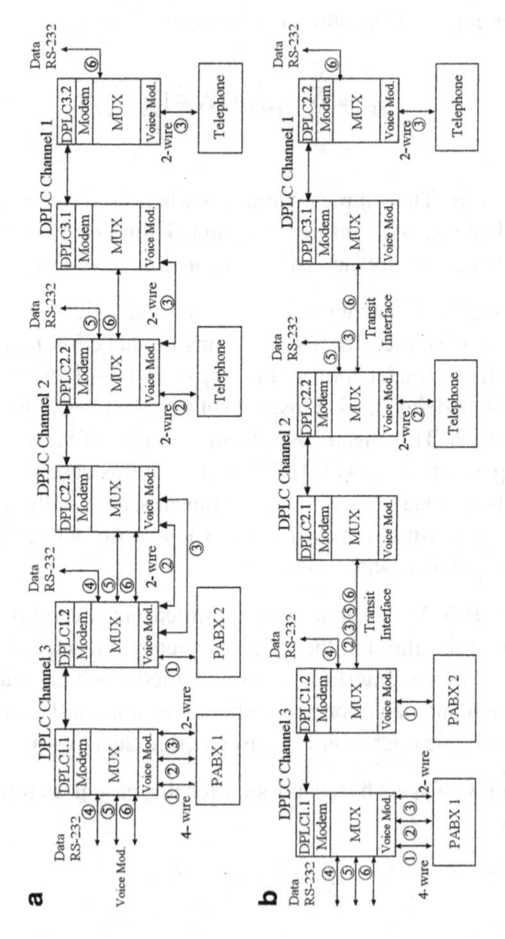

Fig. 5.10 Schemes of DPLC channels with transit sections

three G.729 retransmissions can be used for technological voice channels. When using DPLC equipment with a latency of 40 ms, such systems exist in practice; the number of transit sections for G.729 speech codec can be increased to 5 or 6.

Packet DPLC channels provides their own transit options.

Variant 3, Fig. 5.11a, b Voice signal is transmitted via DPLC channel using VoIP technology, for which the issue of transmission delay calculating has already been considered (see Sect. 5.4.3). The situation is complicated by the fact that, when evaluating delay in a channel with transit sections, it should be taken into account that voice packet in each transit router can be queued for service behind a data or speech packet of parallel sessions.

In this case, in comparison with previous transit schemes, delay depends on the bit rate of DPLC channel in each transit section:

$$t_{\text{EED}} = \sum_{i=1}^{N-1} t_{\text{DPLC}i} + \sum_{i=1}^{N-1} \left(n_i t_{\text{serial.V}i} + t_{\text{serial.}Di} \right) + (q+1) t_{\text{frame}} + t_{\text{alg}}$$

while expressing $t_{\text{serial.}D}$ and $t_{\text{serial.V}}$ through B_{DPLC}, we obtain:

$$t_{EED} = \sum_{i=1}^{N-1} t_{\text{DPLC}i} + \sum_{i=1}^{N-1} n_i \frac{8S_{\text{L2.V}}}{B_{\text{DPLC}i}} + \frac{8S_{\text{L2.D}}}{B_{\text{dPLC}i}} + (q+1) t_{\text{frame}} + t_{alg}$$

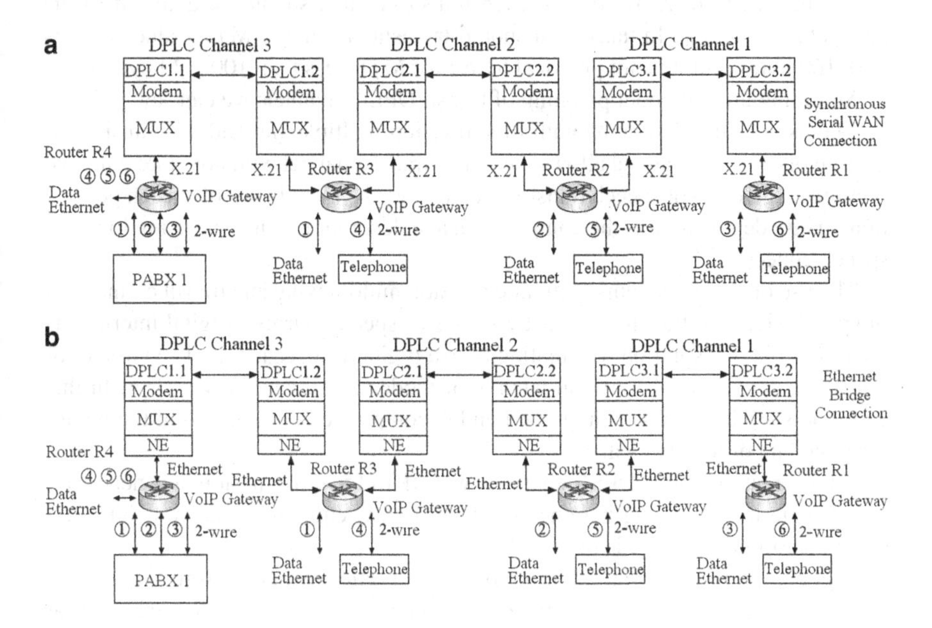

Fig. 5.11 Schemes of IP-DPLC channels with transit sections

where $S_{L2.V}$ – size of the 2-nd layer frame with VoIP packet, bytes; $S_{L2.D}$ – size of the 2-nd layer information frame with data packet, bytes.

Since there are no VF transit connections in IP DPLC channels, the main factor t_{EED} is the delay of VoIP packets.

Let us assume that voice and data signals from three remote substations are simultaneously transmitted via the PLC channel, scheme of which is shown in Fig. 5.11, to the dispatch center (the signals are described in Sect. 5.3.3). DPLC channel bit rates are $B_{DPLC1} = 24$ kbps, $B_{DPLC2} = 48$ kbps, and $B_{DPLC3} = 64$ kbps, Latency of each section of the DPLC equipment is 60 ms. Accordingly, total DPLC equipment delay in the circuit is $3 \cdot 60 = 180$ ms. Speech codecs delay is 45 ms. Total servicing delay for voice packet in transit stations consists of two components: delay of data packets serialization $8 \cdot 215/24 + 8 \cdot 215/48 + 8 \cdot 215/64 \approx 134$ ms, and delays of serialization of parallel session voice packets $8 \cdot 1 \cdot 50/48 + 8 \cdot 2 \cdot 50/64 \approx 21$ ms. As a result, the size of jitter compensation buffer for incoming speech packets should be at least 155 ms.

It is obvious that serialization delay of data packets has a strong negative impact on speech packet transmission time delay. Final $t_{EED(3)}$ value is 380 ms that is inaccessible for all types of voice channels. Even if we apply a DPLC system with a latency of 40 ms, we get $t_{EED(3)} = 320$ ms that will remain inaccessible. Without transmitting data packets at $t_{proc} = 60$ ms $t_{EED(3)}$ will be 246 ms, at $t_{DPLC} = 40$ ms $t_{EED(3)}$ will decrease to 186 ms. Both values are on the border of the permissible for technological and dispatch voice channels.

To meet the practical requirements to limits of t_{EED} for simultaneous transmission of speech and data packets, it is necessary to use DPLC equipment with the lowest latency, and DPLC bit rate in each transit channel should be adjusted to the accessible value t_{EED}. Initially, you should try using a smaller MTU value, such as 140–160 bytes (but it is not practical to use for MTU less than 100–120 bytes).

Summing up said about planning of transit DPLC channels we can say:

The use of DPLC systems with time division multiplexing and VF transit connection of the speech signal allows creating a transit channel with not more than two sections, regardless of speech transmission delay. The main factor of speech perception quality degradation is the compression and decompression cycle using low-rate speech codecs.

The use of DPLC systems with time division multiplexing and transit connection of speech signal in transit stations by means of special-purpose digital interface is the most effective solution for creating transit digital PLC channels. The number of transit sections depends only on latency of modems and can reach 5 or 6. In this case, the speech signal is compressed and decompressed only once – at the beginning and end of the channel.

When creating transit DPLC channels with IP traffic transmission, it is necessary to calculate the following parameters of the DPLC equipment (latency, bit rate), when condition $t_{EED} < t_{max}$ is met.

A good option to increase number of transit sections when transmitting speech signals is to partially process and transit them in APLC mode, for example, at remote sections of transit channel. In this case, additional delay of speech transmission does not exceed 5... 15 ms. To avoid echo signals, the transit of the speech signal should be performed via 4-wire E&M interfaces. In these sections, data should be transmitted in parallel with speech using QAM or OFDM modems operating in the remaining part of the operating frequency band (see Fig. 1.3c).

5.6 Providing Reliability of DPLC Channels when Selecting Operating Frequencies

Required DPLC channel reliability (ChR) (for any other PLC channels as well) should be ensured during creation of these channels at all stages: planning, commissioning, and, finally, during the operation itself.

In this section, we will consider the issues of reliability ensuring only while planning work. In this case, both the work performed at the initial stage (calculation of the maximum allowable channel frequency) and at the last stage (when final calculation of the channel parameters for selected frequencies and operation scheme with proving its operability) are considered. Correctness of the solutions laid down at these stages (especially at the second stage), to a large extent, ensure reliable operation of the channel after its commissioning.

As for other, also very important stages of the channel life, here we note only the need to carry out full volume of required by relevant standard measurements of main parameters of the HF path and the DPLC channel when putting the channel into operation. The channel should be put into operation only if these parameters comply with existing standards. As for the measurement methods, we refer to [11].

When discussing ChR, we consider only those aspects which are not defined by quality of the equipment and relate to parameters of the HF path attenuation and wideband disturbances on the output of this HF path. In this context, the issues of ensuring the ChR of DPLC channels are reduced to ensuring their noise immunity under specified HF path conditions.

One of the main parameters of the DPLC channel which defines its noise immunity is minimum permissible SNR at the receiver input, which still ensures that the required BER is met at a given DPLC bit rate.

To define noise immunity of the DPLC channel, it is necessary to:

- Correctly define conditions under which the noise level should be calculated, taking into account statistically distributed noise parameters discussed in Chap. 4.
- Consider disturbances influence on noise immunity of the PLC channel (wideband disturbances due to corona noise discussed in Chap. 4).

Table 5.6 K_{OPL} values for different voltage OPLs

OPL voltage, kV	35	110	220	330 and more
K_{OPL} value, dB	0	2	6	9

- Correctly select coupling scheme to the overhead line and to calculate attenuation of the HF path which, at a known transmitted signal level, defines its reception level.

Below, we will consider recommendations given in [12] for calculating interference and attenuation of HF paths and will show basic conditions under which the planning channel will work after implementation.

In accordance with given in [12] methodology for defining maximum allowable frequency of the PLC channel and for final definition of the projected channel parameters after selecting the frequencies, the following parameters should be calculated:

- Overlapped attenuation for modulated HF signal
- Standardized attenuation margin which considers possible increase of the path attenuation relative to its calculated value and noise level used for calculation of the overlapped attenuation
- Attenuation of the HF path and its non-uniformity (calculation is carried out without influence of IHRD on the overhead line conductors)
- Calculated margin for attenuation obtained as the difference between overlapped attenuation and attenuation of the HF path

Remind that maximum allowable frequency is the highest frequency in PLC spectrum, which provides required bit rate and ChR for PLC channel under consideration (with specified OPL parameters, corona noise, weather conditions). For example, PLC spectrum is up to 500 kHz, but result of the calculation shows that the maximum frequency for considered PLC channel cannot be higher than 350 kHz. If selected operating frequency will be higher than 350 kHz, this PLC channel will be inoperable during adverse weather periods.

The channel is considered operable if calculated attenuation margin is not less than its required value, and non-uniformity in the HF path attenuation corresponds to the required value.

5.6.1 Calculation of the PLC Channel Overlapped Attenuation

Formula for overlapped attenuation can be written as:

$$a_{\text{overl.}} = p_{\text{TX}} - p_{\text{RXmin}} \tag{5.1}$$

where p_{TX} – level of modulated HF signal transmission (defined by PAPR value of the signal power); $p_{\text{RX.min}}$ – minimum acceptable level of received modulated signal which, at a given level of noise, still provides reception with the required confidence (bit error rate).

Minimum acceptable reception level is defined as:

$$p_{\mathrm{RXmin}} = p_{\mathrm{noise}} + 10\log \Delta f + SNR_{\mathrm{min}} \tag{5.2}$$

where p_{noise} is standard level of corona noise in 1 kHz band, dBm; Δf is frequency band for which noise is defined, equal to the operating band of DPLC equipment modem, kHz; SNR_{min} minimal value of signal-to-noise ratio for a given bit rate. Required SNR is defined by characteristics of the DPLC equipment modem of a particular manufacturer.

Data on the standard value of the corona noise level for overhead lines of different voltage classes are given in Chap. 4 (see Table. 4.10).

The main task in calculating according to (5.2) is to define calculated level of corona noise and minimum permissible SNR value.

Minimum permissible SNR value specified by manufacturer is given for white noise disturbances. This is because all certified noise generators used to verify compliance with the requirements of the technical specifications for the DPLC equipment produce white noise disturbances.

At the same time, studies [13, 14] have shown that if in the formula (5.2) for calculated noise level due to corona p_{noise} average RMS level (see Fig. 4.47) is used, then, to ensure that the bit error rate is not more than the standardized one (for example, 10^{-6}), SNR for corona noise should be increased relative to that specified by manufacturer a value up to 6... 9 dB. The conclusions and recommendations have been fully confirmed by the experience of DPLC channel operation.

Thus, to ensure necessary channel performance, in the calculation formula (5.2) for calculating minimum permissible reception level, correction to the SNR value (corona factor K_{OPL}) should be made. In this case, the formula (5.2) is written as:

$$p_{\mathrm{RXmin}} = p_{\mathrm{noise}} + 10\log \Delta f + SNR_{\mathrm{min}} + K_{\mathrm{OPL}}$$

Corona factor K_{OPL} values are given in Table 5.6.

Calculation of overlapped attenuation for DPLC systems operating at adjacent receiving and transmitting frequencies also has some features. No matter how powerful is echo canceller in the receiver and how perfect is PLC equipment, these devices cannot completely exclude influence of the own transmitter on the own receiver if their frequency bands are adjacent. In the practice of PLC, manufacturers impose restrictions on the use of DPLC channels at adjacent frequencies that consist in limiting overlapped attenuation value. The manufacturer should provide the planner with information about these restrictions. In addition, for a variant with separated transmission/reception frequency bands, manufacturer should specify minimum necessary gap between these frequencies bands in which the above restrictions may not be taken into account.

5.6.2 Required Attenuation Margin

To define the channel operability (at known calculated attenuation of the HF path and known DPLC system), it is necessary to compare calculated attenuation of the HF path for this channel with equipment overlapped attenuation (channel budget).

In this comparison, it is necessary to think the fact that conditions under which the channel should work may differ from those under which both calculated attenuation of the path and equipment-overlapped attenuation were defined.

Taking the said into account, calculated attenuation of the HF path which can guarantee operation of the channel with required noise immunity should satisfy the inequality:

$$p_{\mathrm{TX}} - a_{\mathrm{marg1}} \geq a_{\mathrm{HFpath}} + a_{\mathrm{marg2}} \tag{5.3}$$

in which a_{marg1} is the value which determines reduction of the overlapped attenuation due to reasons not considered in the calculation according to (5.2). For DPLC channels, this reason is increase of corona noise level above calculated (50%) value at adverse weather conditions; a_{marg2} – value which defines the possible increase of the path attenuation in comparison with calculated one. For DPLC channels, this is attenuation due to ice coating.

Usually, in the formula (5.2), values a_{marg1} and a_{marg2} are combined, and this formula is represented as:

$$p_{\mathrm{TX}} \geq a_{\mathrm{HFpath}} + a_{\mathrm{marg}} \tag{5.4}$$

In this formula:

- Overlapped attenuation is defined for a noise level for which probability of exceeding the level is 50% (it can be defined using both exact calculation methods and the table).
- The HF path attenuation a_{HFpath} is defined without accounting of IHRD effect. Calculation of the required attenuation margin which ensures channel performance in adverse weather conditions will be considered separately.

It should be noted that the term "attenuation margin" does not exactly correspond to what is essence for DPLC channels. It is more correct to call this parameter "reserve for possible degradation of HF path parameters defining channel performance." But this is the established term which is used to describe not only DPLC channels but also all kinds of PLC channels including channels for teleprotection, and in this text, we will use the term "attenuation margin." However, we will keep

Table 5.7 Values of channel non-operability time for different x values, %

x, %	2.5	1	0.25
$\Delta p_{\mathrm{interf}}$, dB	9	11.2	13
Time during which the channel is out of operation, hours per year	219	87.6	21.9

in mind that this term in its general sense refers to description of the channel's margin for reducing SNR (a_{marg1} – reducing SNR due to increase of corona noise level and a_{marg2} – reducing SNR due to increase of a_{HFpath} due to ice coating).

The attenuation margin value in (5.4) is defined in general form as:

$$a_{\text{marg}} = a_{\text{marg1}} + a_{\text{marg2}} = \Delta p_{\text{noise}} + \Delta a_{\text{IHRD}} \tag{5.5}$$

When defining minimum required margin for analog and digital channels for speech and data transmission, it is assumed that probability of coincidence in the time of increasing corona noise level and increasing attenuation from IHRD can be ignored.

Taking this assumption into account, only one value (the largest) of the two values (Δp_{noise} and Δa_{IHRD}) should remain in the formula (5.5).

Therefore, it is recommended to define the standardized margin value a_{marg} as:

$$a_{\text{marg}} = \Delta a_{\text{IHRD}(f)} \tag{5.6}$$

(but not less than Δp_{noise}, dB), where Δp_{noise} is the required correction which takes into account increase of noise level above its 50% value (defined taking into account some considerations which will be discussed below).

Special case of formula (5.6) is used in practice when planning DPLC channel:

$$a_{\text{marg}} = \Delta a_{\text{IHRD}(f)} \tag{5.7}$$

(not less than 9.0 dB)

When calculating the margin for analog and digital voice and data channels, it is assumed that used Δp_{noise} value is equal to such its value for which probability of exceeding this value is equal to 2.5%. According to Fig. 4.45, this value Δp_{noise} is 9 dB.

It follows from the formula (5.6) that the required margin is defined by the largest term in (5.5) Δa_{IHRD} or Δp_{noise}.

Considering the above, standardized value of the attenuation margin for DPLC channels in the formula (5.6) should ensure the channel performance both at increase of the path attenuation due to ice (Δa_{IHRD}), and at increase of corona noise level due to adverse weather conditions (Δp_{noise}).

We will show the basis for defining Δp_{noise} value necessary for calculating margin for attenuation according to (5.6).

To begin with, we emphasize that, as can be seen from formula (5.6), specified value of Δp_{noise} affects the used attenuation margin value (and through it, result of calculating the maximum allowable channel frequency) only if

$$\Delta p_{\text{noise}} = \Delta a_{\text{IHRD}}$$

Considerations for the method of selecting Δp_{noise} value are illustrated using an idealized channel model in which HF path parameters (including attenuation) are invariable over time. The only variable in this model is noise level at the receiver input. At that, it is assumed that this noise voltage is a random variable distributed according to a normal law (at that, noise level is distributed according to a logarithmically normal law). The integral distribution function of this noise level corresponds to Fig. 4.45 (i.e. to statistically distributed characteristic of actual overhead line corona noise).

In this DPLC channel model, SNR at the channel receiver input changes only due to changing in time the noise level that occurs randomly in accordance with integral distribution function of noise shown in Fig. 4.45.

Let us consider conditions for maintaining the channel performance when the receiver is affected by noise with the parameters described above.

To describe these conditions, we introduce a concept of "channel noise efficiency factor" (K_{noise}) (hereinafter, for brevity, we will call it "noise factor").

Let

- Noise level $p_{noise} = p_{50\%} + \Delta p_{noise}$ at the receiver input is such that SNR at the receiver input ensures the channel performance with given bit rate and BER;
- Probability of exceeding this level, as defined in Fig. 4.45, is x (%).

In this case, the noise factor is defined by the formula

$$k_{noise} = (1 - x/100)$$

where x, %, is probability of exceeding the accepted noise level $p_{noise} = p_{50\%} + \Delta p_{noise}$.

For the channel model used, x value, %, corresponds to time (as a percentage of total channel operation time) during which the channel will not be operable.

In Table 5.7, some values of Δpnoise and corresponding probability of exceeding this level x, %, are given which were obtained from the data used for drawing the graph in Fig. 4.45. The same table shows number of hours per year during which the channel operating under above conditions will lose operability.

Despite some ideality of the channel model for which the data given in Table 5.7 are obtained, this data can be a basis for estimating required Δp_{noise} value in (5.6) and duration of non-operability corresponding to this value.

Use of data obtained for the "ideal" channel model, when considering a DPLC channel operating along real overhead lines, is quite acceptable for time interval within which there are no conditions which lead to IHRD forming (including snow sticking) on the phase conductors of the overhead lines. As a rule, number of days in this period does not differ much from total number of days in the year.

Having in mind the said above, we can recommend to select values of Δp_{noise} and x, %, taking into account described considerations and customer's channel reliability requirements.

5.6.3 Application of Accurate Calculation Methods for HF Path Parameters

Above, the specificity of HF path parameters and overhead line noise were considered when overlapped attenuation of DPLC equipment and attenuation margin were determined. Taking these specificities into account, it is possible to improve the performance of the DPLC channel.

In this section, we will consider recommendations to address the same objective of improving the DPLC channel performance when calculating of HF path attenuation and non-uniformity of this attenuation carried out during the planning at stage of final calculation of the DPLC channel parameters.

Requirements for the HF path parameters, which are necessary to ensure DPLC channel performance, are significantly more severe than for APLC channels. This is determined by the fact that much more complex modulation and synchronization principle are used in DPLC channels than in APLC channels.

As a result, operability of the designed DPLC channel after its implementation is largely determined by degree of adequacy of the HF path parameter calculation results obtained during planning and which were used as a basis for making conclusion on the channel conformity with standard requirements, and actual values of these parameters measured during input of the channel into operation. To ensure necessary accuracy of the calculation, let us consider conditions which should be taken into account when calculating attenuation of the HF path.

When having in mind:

- Complexity of processes in the HF path when it transmits a HF signal over non-homogeneous overhead lines shown in Sect. 4.1
- Necessity of accurate calculation to define HF path attenuation

At final calculation of the channel parameters at selected frequencies, it is necessary to apply accurate calculation methods using WinTrakt software. At that, the fact should also take into account that there are no simplified methods for calculating return loss being an important parameter for DPLC channels and that this parameter, along with HF path attenuation, allows complete evaluation of the DPLC channel performance.

The practice of using simplified calculation methods has remained from the time when computers were not widespread and the use of software for calculations was practically excluded.

Currently, when a personal computer is available to every specialist and there are proven programs which provide accurate methods for calculating the HF paths parameters and disturbances due to corona noise, the application of accurate calculation methods, in principle, is not limited.

Few general words about accurate calculation methods and reliability of the calculation results.

First, let us define a concept of "accurate calculation method." Any calculation method is based on a system of certain assumptions which simplify the described actual model and can be called "accurate" only conditionally. In relation to calculation of HF path parameters and wideband disturbances in these paths, we will call accurate the method based on theoretical provisions describing the processes in existing paths with the greatest degree of approximation to reality. We will assume that, at present, these provisions are formulated to the greatest extent in [11]. WinTrakt and WinNoise software implements algorithms based on the calculation methods described in this book.

When using the calculation results, it should always be remembered that the HF path and corona noise parameters obtained in these calculations will correspond to actual parameters only to the extent to which the calculated HF path scheme and initial data used in the calculation scheme correspond to reality.

This simple thought is not always taken into account. It is believed that the very use of software is already a guarantee of the calculation result reliability.

Therefore, when using software for calculating the HF paths parameters and interference due to corona, special attention should be paid to obtaining and entering up-to-date and reliable information on all elements of the HF path scheme. It should be kept in mind that source data may be incomplete and may contain errors.

To be able to analyze the obtained source data of a design scheme and, if necessary, to adjust them, engineering specialists should be familiar with the available literature and should have an idea of possible values of those parameters of high-voltage equipment which affect the HF path parameters. This knowledge makes it possible to evaluate the initial data obtained for the design and, if necessary, to seek their clarification.

Examples of such knowledge can be competence in principles of selection of scheme for coupling to the line, about a system for hanging and earthing steel and conductive groundwires and optical groundwires (OPGW), about equivalent circuit scheme for high-voltage equipment in the field of high frequencies, etc.

It should be noted that, when using accurate calculation methods (performing calculations using the WinTrakt software), calculation should be performed within the frequency range of line traps and coupling devices (line matching units) and

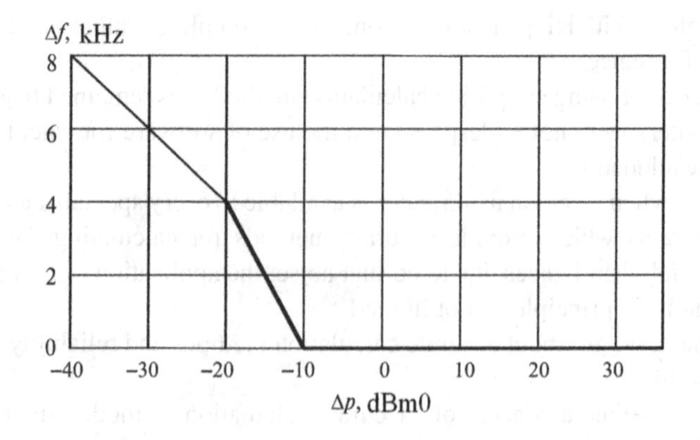

Fig. 5.12 Graph of required frequency gap Δf versus level difference Δp

should perform analysis of the obtained frequency response characteristics. If calculation shows presence in the frequency response of an attenuation pole with significant attenuation increase, it is necessary to consider the need to take measures to change the coupling scheme in such a way that the attenuation pole is not be present or nominal transmission/reception frequency bands of the channel are located far away from this pole in the lower-frequency range.

In conclusion, we should mention those cases in which simplified calculation methods are used in designing digital PLC channels. These methods are used for:

- Defining maximum allowable frequency of the channel. At this stage of planning work, it is quite acceptable to use approximate calculation methods, since the calculation results are used as a top estimate of the maximum frequency, rather than its accurate value. Accurate methods are recommended only in the cases when maximum allowable channel frequency defined using simplified calculation methods is so low that it is impossible to further select transmission/reception frequency bands.
- When defining the transmitters, signal levels at the installation sites are affected by receivers, in task of selecting frequencies for PLC channels operating in the common electric grid.

5.6.4 Providing EMC of APLC and DPLC Channels Operating in One Electric Grid

Electromagnetic compatibility (EMC) of the PLC channels operating in the same electric grid is provided by appropriate frequency planning for these channels. There is simplified method for frequency planning procedure where repetition of same frequencies is possible after specified number of lines and substations between channels [12]. In accurate method, mutual interference of all PLC channels operating in the same electric grid takes into account. This method is described in [12] and is used every time when new channel is planned and operating frequency bands for the channel, which provides required EMC with all existing and previously planned channels, are selected.

Here we will not entirely consider this procedure; we will only say that the procedure is applied every time when frequencies for a new channel are selected and consist of the following steps:

1. Evaluation (calculation) of output power by transmitters at input of the receiver of the considered channel, for all existing channels available in the electric grid
2. Evaluation (calculation) of power output by the transmitter of the considered channel at the inputs of all receivers of all existing channels available in the electric grid
3. Definition of frequency bands in which the levels of mutual interference of new channel and all other channels calculated at the first two steps are within the

established limits which ensure performance of all channels (their operation with the specified parameters). Performances of channels involved in frequency planning for a new channel are defined in such a way that, on the one hand, the time required for correct frequency selection is not increased too much, and, on the other hand, mutual effect of all channels with influence levels exceeding specified values is taken into account. As a rule, all channels located at substations separated from the end substation of the new channel by at least three power lines should be considered.

This calculation should be made by collecting data on selectivity of all receivers involved in the consideration and data on permissible level of interference caused by each transmitter of the affected channels for which frequency bands coincide with the frequency band of the affected receiver of the channel in question.

The above steps for selecting frequencies are often rather complex and voluminous task. This is due to the complex topology of high-voltage grids with many electric connections between substations, i.e., signal propagation paths between affecting transmitter and affected receiver. The level should be defined for the path with the lowest attenuation. In addition, the number of channels involved in the PLC channel consideration is also large that significantly increases number of considered options of influence ways.

Of the above steps for selecting frequencies (providing required EMC of PLC channels), we will consider only part of the third step, namely, principles used to define minimum allowable gap of the nominal frequency bands of channels for given affecting signal level at the receiver input.

Minimum required frequency gap between the edges of the nominal frequency bands of the affected transmitter and affected receiver (Δf) is defined using the typical graph shown in Fig. 5.12.

Note: when drawing Fig. 5.12, it was assumed that maximum value Δp at $\Delta f = 0$ (in this case 35 dBm0) corresponds to the Δp value at which frequency repetition is possible; minimum value of Δp at $\Delta f = 0$ (in this case −10 dBm0) corresponds to the Δp value at which selection of adjacent frequency bands is possible.

In the graph (Fig. 5.12), for combined APLC and DPLC channels, Δp value which is used to calculate minimum required frequency gap (Δf) is defined as the difference between minimum receiving level and level of interfering (affecting) signal $p_{\text{interf.}}$ defined on the first two calculation steps:

$$\Delta p = p_{\text{RXmin}} - p_{\text{interf.}}$$

Data for drawing the graphs of the type shown in Fig. 5.12 are provided by the equipment manufacturers in accordance with the results of the equipment measurements carried out using the method described below.

According to this method, acceptable Δp value is defined as the difference between the level of the DPLC modulated signal transmitted during testing with a required bit rate and the level of interfering signal at the input of the channel receiver at which BER does not exceed 10^{-7}.

In this test, interfering sinusoidal signal should have a frequency of:

- At measurements beyond the nominal DPLC frequency band: frequency spaced from the edge of the nominal receiving frequency band by 100, 4000, and 8000 Hz.
- At measurements within the nominal frequency band: frequency changing within the rated receiving band.
- Features of graph shown in Fig. 5.12 based on the measurement results are as follows:
- Result of measurements within the receiver nominal frequency band is used as Δp value for the rightmost point of the x axis at $\Delta f = 0$ (frequency repetition is possible).
- Result of measurements at a frequency which is separated from the edge of the nominal receiving frequency band by 100 Hz is used as Δp value for the leftmost point of the x axis at $\Delta f = 0$ (adjacent bands are possible).

5.7 Design of DPLC Channels with Bit Rate Adaptation

In this section, we will discuss the use of DPLC channels with bit rate adaptation, specifics of determining maximum allowed frequency ($f_{\text{max.chan}}$) for these channels compared to DPLC channels without adaptation, final calculation of DPLC channel parameters with bit rate adaptation, examples of long-term monitoring of DPLC channel HF path attenuation, noise, and SNR taking into account weather factors.

5.7.1 Purposes of Use of DPLC Channels with Bit Rate Adaptation

Let us define in which cases it is advisable to use the DPLC channels with dynamically adaptive bit rate and in which cases the best solution is use of a fixed bit rate DPLC channel.

Earlier, in Table 5.1, information signals transmitted in the communication networks of electric power companies were considered. Signals related to dispatching management: dispatch voice, telecontrol, and power metering systems have high and middle priorities and should be transmitted under any operating conditions of the communication channel. Technological voice, transmission of data for access to substation devices for reading event recorders, Internet, etc. belong to the signals of normal and low priority, and in the case of adverse weather conditions, their interruption is allowed.

The following rules for using DPLC channels with bit rate adaptation can be formulated:

Rule 1 Different priority data, such as dispatcher and technological voice, are transmitted by the same DPLC channel. At that, disabling technological voice when the DPLC channel bit rate decreases due to adverse weather conditions, such as ice, is allowed.

This does not apply to replacing high-priority signals with low priority ones when dividing available channel capacity (see Rule 4). When under normal operating conditions, the entire channel is busy, for example, by low-priority data transfer services, but a voice signal occurs, required bit rate is immediately allocated to it.

Let a DPLC channel connects two PABX via six trunks with E&M signaling. One trunk is used for dispatcher voice, 5 others for technological voice, and 3 of them are used for office conversations. It is extremely important to ensure operation of only one trunk – for dispatcher center. For the other trunks, under adverse conditions, unavailability can be assumed: first, disabling office trunks, and then, if necessary, technological voice.

More complex is another case, when transit DPLC channels connect several substations, and a direct channel of dispatcher voice and dial-up channels of technological communication are organized for each of them, either to the central PABX or between PABXs located at the substations. The determining factor here is mutual geographical location of the substations. Since there should always be enough resources for the system/transit to transmit dispatcher voice signals, then, if a substation is located sequentially/linearly on the last transit section, resources should be available for operation of three dispatcher voice channels. And adaptation possibilities here are minimal compared to the DPLC channel at the beginning of the circuit/transit section where it can disable several channels of technological voice.

Voice signal prioritization function is implemented either in the settings of the built-in DPLC equipment multiplexer or in an external multiplexer. The multiplexer should work with the DPLC equipment modem at synchronous mode. The modem is a master device, and it sets synchronization rate, while the multiplexer is a slave device. Described speech transmission scenarios relate to the DPLC channels discussed in Sects. 5.4.1 and 5.4.2. If data are transmitted via RS-232 interface in the DPLC channel (see Sect. 5.4.1), they are, as a rule, data of telecontrol or power metering signals, and required modem bit rate under adverse conditions should be enough to transmit all data signals, including transit ones.

Rule 2 In the DPLC, channel resources are divided between time division multiplexing and packet switching subsystems. In this case, a packet data transmission using IP network protocols is applied, for example, transmission of data of telecontrol or power metering systems based on IEC 60870-5-104 protocol, data of various devices monitoring via SNMP, reading recorders, etc. Speech is transmitted via the built-in multiplexers, and prioritization of voice signals is performed as described in Rule 1. Bit rate of the DPLC modem operation under adverse conditions should be enough for transmitting dispatcher voice signals and high-priority data from telecontrol system.

Power metering data are not critical to the delivery delay and can be in turn transmitted from all polling points.

Low-priority signals can be transmitted at good weather or if there are free resources in adverse weather, for example, when ice coating. Issues of such DPLC channel creation are discussed in Sect. 5.4.2.

Rule 3 In a converged DPLC channel with IP-based packet speech and data transmission (see Sect. 5.4.3), signals are controlled by a combination of the approaches described in Rules 1 and 2. The modem bit rate under adverse conditions should be enough to transmit signals of dispatcher voice, telecontrol, and power metering systems.

Prioritization/division of packet traffic is performed by external devices – network switches and routers – in accordance with the principles discussed in Sects. 3.2.2 and 5.4.3.

Rule 4 In some cases, when the solutions are being agreed with the operating electricity company, data signal transmission with the use of resources intended for voice can be allowed if, of course, the resources are free of voice connection. An example of this is the PLC channel used to transmit dispatcher voice signal and data signals of telecontrol system. This approach can be implemented for any variant of DPLC channel structure (see Sects. 5.4.1, 5.4.2, and 5.4.3); it is important that the DPLC equipment multiplexer or external network device has necessary functions for prioritizing signals and allocating system resources.

Rule 5 DPLC channels with a fixed bit rate should be used in cases when, according to technical conditions, signals of the same priority are transmitted through the DPLC channel. As a rule, these are high-priority signals and their transmission should be guaranteed under all conditions. The simplest example: a DPLC channel is used for transmitting a single dispatcher voice signal and a data signal of the telecontrol system.

However, this rule should not limit the ability to adapt the channel and disable one or more high-priority data signals to try to safe voice communication in the condition of extreme ice or a line failure, such as a phase conductor breakage.

5.7.2 Calculation of the PLC Channel Maximum Allowable Frequency

Selection of any type of PLC channels operating frequencies (including DPLC channels) should be carried out in two stages. At the first stage, maximum allowed frequency of the channel is defined. At the second stage which can significantly differ in time from the first, nominal channel transmission/receiving frequency bands are selected, and the HF path coupling scheme is finally formed; final calculation of the channel parameters is performed. This calculation should confirm the channel operability (note among these parameters attenuation margin and non-uniformity of the HF path attenuation).

A method for determining maximum allowable frequency and a method for conducting final calculation of parameters of the DPLC channel with bit rate adaptation to changing SNR have some differences compared to the corresponding methods used for the DPLC channel without adaptation.

So, for example, in the case of designing a DPLC channel without bit rate adaptation, maximum allowable frequency of the channel is defined uniquely taking into account ensuring standardized minimum allowable attenuation margin. The attenuation margin defined at the stage of PLC channel final calculation is also defined uniquely by the value of attenuation overlapped by the equipment and calculated attenuation of the HF path (according to formula which can be obtained from (5.4)):

$$a_{\text{marg}} = a_{\text{overl.}} - a_{\text{HFpath}} \tag{5.8}$$

In the case of designing a PLC channel with bit rate adaptation, maximum allowable channel frequency depends in the principle of the bit rate (steps of adaptation) for which it is defined. That is, when defining this frequency, solution is multivalued, and some addition to methods for determining maximum allowable frequency which eliminates this ambiguity (maximum used channel frequency should be the same for rates of all adaptation steps in the considered channel) is necessary. Also, an addition to the stage of verification of final calculation for the channel with the selected nominal frequency bands is required, which should include a method for determining actual attenuation margin which the channel will have at each of the adaptation steps.

The content and essence of the above additions depend on the chosen scenario defining principles for selection of conditions for transmitting modulated HF signal at each step of adaptation. To date, there is no sufficient consideration of this issue providing necessary recommendations.

To fill this gap, we present recommendations for solving this problem for a special case of implementing the rules for using DPLC channels with bit rate adaptation described above.

The special case is that:

- At the first step (it is allowed at the first and second step as well), the channel can be interrupted when IHRD appears on the overhead line phase conductors.
- At subsequent steps (usually from the second one), the channel operability should be ensured at increasing the HF path attenuation due to IHRD on the overhead wires.

Stage 1 Calculation of maximum allowable frequency of the channel. Algorithm for calculating maximum allowable frequency consists of the following steps:

1. Collection of input data. The volume and content of these input data mainly correspond to those set for the PLC channel without rate adaptation. The difference

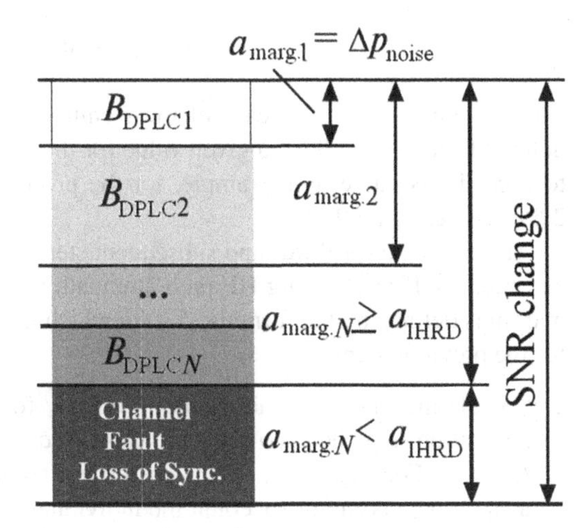

Fig. 5.13 Adaptation steps of the DPLC modem

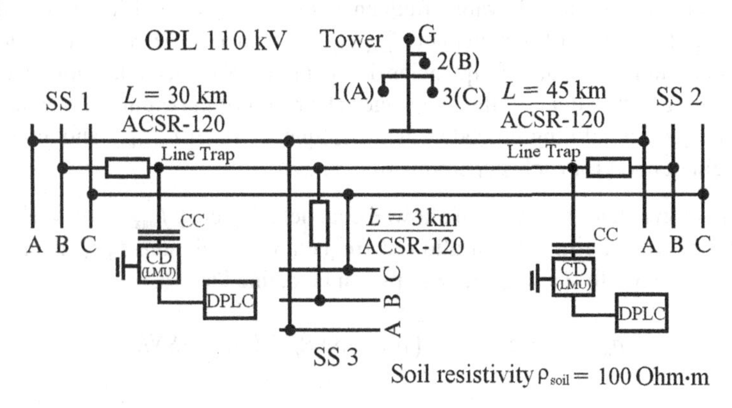

Fig. 5.14 OPL scheme and initial data for the line

is that addition of specific initial data related to the implementation of the adaptation procedure is necessary. Together with these additional data, required number of adaptation steps should be set, and, for each adaptation step: bit rate, modulated HF signal transmission level, required SNR, and requirements for conditions under which the channel should remain operable.

The term "conditions under which the channel should remain operable" means specified parameters of IHRD (type of IHRD, ice layer thickness, length of section with IHRD) and maximum noise level of the channel at a given step of adaptation, at which the channel should remain operable.

Actually, this is a set of requirements for standardized value of the attenuation margin at each adaptation step.

As already mentioned above, it is generally assumed that, at the first step, interruption of the low priority services is allowed when IHRD appears on the overhead line wires. In this case, requirement to the attenuation margin for the first step is defined according to the formula:

$$a_{\text{marg.req}} = \Delta p_{\text{noise}} \tag{5.9}$$

This formula is a special case of the formula (5.6). Value of Δpnoise should be defined from Fig. 4.45 for a given value for the probability equal to the probability to exceed this value. For example, for the probability to exceeding this value by 2.5%, i.e., $\Delta p_{\text{noise}} = 9$ dB.

At the second (or third) and subsequent steps, the channel operability should be ensured at IHRD affecting HF path attenuation. In this case, required attenuation margin is defined by the formula (5.6) in which Δp_{noise} is assumed to be the same as for the previous steps.

2. Calculating maximum allowable frequency for each adaptation step. For each adaptation step, the maximum allowable frequency should be defined ($f_{\max 1}$, $f_{\max 2}$, $f_{\max 3}$, ...). For each step, this is performed similarly to DPLC channel without adaptation, taking into account the individual attenuation margin requirements.
3. Defining maximum allowable frequency of the channel. Maximum allowed channel frequency of the channels $f_{\text{max.chan}}$ is assumed to be equal to the least of the maximum allowable frequencies defined in p.2 for each adaptation step.
4. Obtaining output data for the first stage. The result of maximum allowable channel frequency defining procedure is the frequency itself $f_{\text{max.chan}}$, and attenuation margin defined for each step at this frequency.

It can be shown that, if maximum allowed channel frequency $f_{\text{max.chan}}$ was selected in accordance with the maximum allowed frequency of k-th step $f_{\max.k}$, then margin obtained for any i-th ($i = 1, 2, 3,...$) steps can be defined as:

$$a_{\text{marg.req}i} = a_{\text{marg.req}k} + \left(p_{\text{TX}i} - SNR_i\right) - \left(p_{\text{TX}k} - SNR_k\right) \tag{5.10}$$

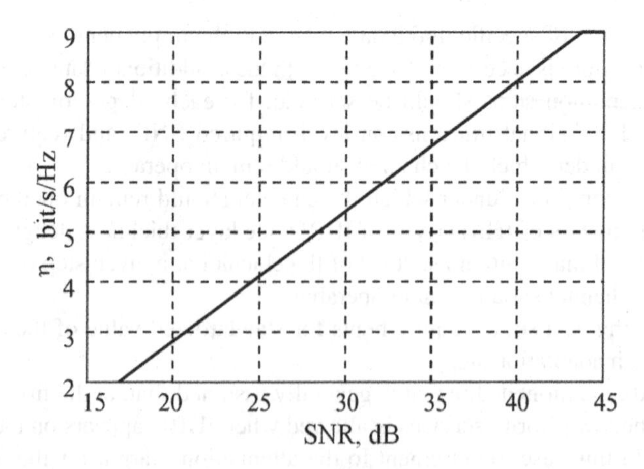

Fig. 5.15 Graph of modem spectral efficiency

where $a_{\text{marg.req}k}$ required attenuation margin for the k-th step which maximum allowed frequency is assumed to be maximum allowed frequency of the channel.

Attenuation margin values for different steps are define using the relation:

$$a_{\text{marg1.}} < a_{\text{marg2}} < a_{\text{marg}N} \tag{5.11}$$

where N is the number of adaptation steps.

Stage 2 Final calculations of the channel parameters

Let us consider here only the part of final calculation which is related to attenuation margin definition.

Attenuation margin is defined for each channel adaptation step. Calculation is performed according to the formula (5.8), in which overlapped attenuation of the equipment is defined for each adaptation step according to the formula (5.1), considering K_{OPL} in (5.2), and attenuation of the HF path is defined for the frequencies corresponding to the nominal receiving frequency bands. Let us recall that these frequency bands in the area with placed upper limit – maximum allowed frequency of the channel. Therefore, these margins will be more than that defined by (5.10) by the difference between the HF path attenuation defined for maximum allowable channel frequency and nominal frequency used. At that, of course, inequality is valid (5.11).

Figure 5.13 shows an example of channel adaptation with N adaptation steps. Each step corresponds to bit rate-B_{DPLC1}, B_{DPLC2},..., $B_{\text{DPLC}N}$. In this example, at the last step of adaptation, the channel operation should be provided with the largest attenuation due to IHRD used at the earlier steps.

To keep the channel operable at some intermediate i-th step, for example, if ice falls, the condition $a_{\text{marg}i} \geq a_{\text{IHRD.req}i}$ should be met at this i-th step. But when $a_{\text{marg}i} < a_{\text{IHRD.req}i}$, then transition to the next $(i + 1)$ adaptation step occurs.

If, for the last step, $a_{\text{marg}N} < a_{\text{IHRD.req}N}$ (that means that the attenuation margin is insufficient for operation at the N-th step of adaptation), then synchronization between modems is lost, and the channel fails.

Here is an example of determining maximum allowable frequency $f_{\text{max.chan}}$ and final parameter calculation for the DPLC channel with bit rat adaptation.

For calculations, we will use information provided in this and previous chapters and reference data from [12].

Initial data and the problem statement. The channel is planned along 110 kV overhead line. Diagram of the line indicating lengths and other data is shown in Fig. 5.14. The channel is located between SS1 and SS2. The branch substation SS2 is not used for communication and is equipped with the line trap installed on the tower, in the coupling phase conductor, and at the tap line location. Coupling scheme – phase 2 (B)-to-earth. Type of phase conductor is ACSR-120 corresponding to aluminum conductor steel reinforced cable (ASCR) with 15.2 mm diameter.

OPL data: all necessary data are shown in Fig. 5.14. OPL is in the third region of ice coating with ice layer thickness of 10 mm.

Data on information transmitted over the channel: two voice channels for dispatcher and technological purposes voice with rates of 7.2 kbps and Ethernet data with a bit rate of 32 kbps should be transmitted over the channel.

The channel should have three steps of adaptation:

The first step with the bit rate B_{DPLC1} equal to 7.2 + 7.2 + 32 = 46.4 kbps;

When SNR decreases, the channel switches to operation with the second step. In this case, technological voice channel is disabled, and Ethernet data bit rate is reduced to 28 kbps, i.e., $B_{DPLC2} = 7.2 + 28 = 35.2$ kbps;

When SNR is further reduced, the channel switches to operation with the third step. At that, Ethernet data bit rate is reduced to 16 kbps, i.e., $B_{DPLC3} = 7.2 + 16 = 23.2$ kbps. Let us round the values: $B_{DPLC1} = 48$ kbps, $B_{DPLC2} = 36$ kbps and $B_{DPLC3} = 24$ kbps.

When defining attenuation margin, Δp_{noise} value should be not less than 9 dB.

Data on DPLC equipment: abstract equipment with typical characteristics:

- Peak envelope power of transmitter PEP = +47 dBm (50 W).
- Level of modulated HF signal. Power allocated to the pilot signal is 1 dB, PAPR of modulated signal is 11 dBm, $p_{TX} = 47 - 1 - 11 = 35$ dBm.
- Nominal frequency band for transmitting/receiving equipment should be to provide average value of modem spectral efficiency in actual DPLC channels equal to 4... 6 bit/s/Hz. The DPLC modem spectral efficiency for AWGN corresponds to the graph shown in Fig. 5.15. Recommended nominal frequency band of the equipment $W = B_{DPLC1}/6 = 8$ kHz. Spectral efficiency at the lowest adaptation step $B_{DPLC3}/W = 3$, whichs provides some margin for operating conditions and reliability of the channel.

Required SNR for white noise disturbance is defined according to Fig. 5.15:

- for $B_{DPLC1} = 48$ kbps $SNR_{min1} = 33$ dB
- for $B_{DPLC2} = 36$ kbps $SNR_{min2} = 25$ dB
- for $B_{DPLC3} = 24$ kbps $SNR_{min3} = 21$ dB

The difference between SNR_{min1} and SNR_{min3} is 12 dB (in actual PLC channels, this value varies more widely).

Next, we will consider calculations for the channel using two calculation methods: simplified and accurate.

For simplified calculation, we assume that, at operation at the first step, the DPLC channel interruption is acceptable when IHRD appears on OPL (standardized attenuation margin is defined by the formula (5.8) considering only noise level increase), and at the second and subsequent steps, the DPLC channel operability is expected to be maintained at IHRD presence (standardized attenuation margin is defined by (5.6)).

Maximum allowed frequencies for the first (f_{max1}), second (f_{max2}), and third (f_{max3}) steps are defined using common methodology applied for DPLC channels without adaptation.

Let us calculate maximum allowable frequency for each adaptation step and define maximum allowable frequency of the channel.

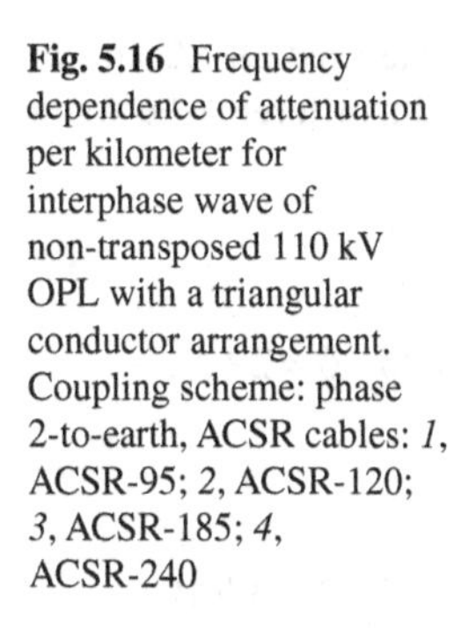

Fig. 5.16 Frequency dependence of attenuation per kilometer for interphase wave of non-transposed 110 kV OPL with a triangular conductor arrangement. Coupling scheme: phase 2-to-earth, ACSR cables: *1*, ACSR-95; *2*, ACSR-120; *3*, ACSR-185; *4*, ACSR-240

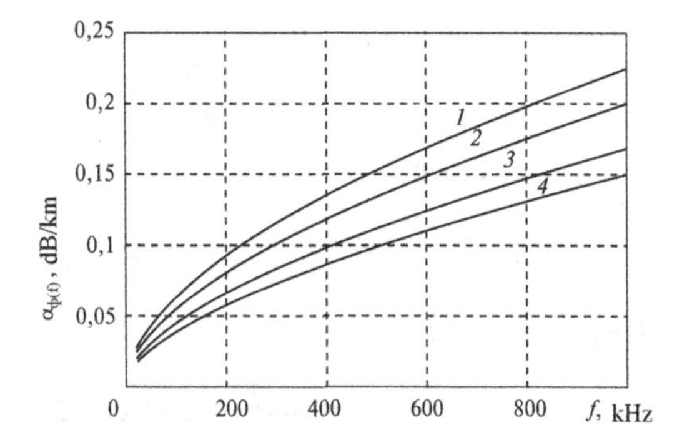

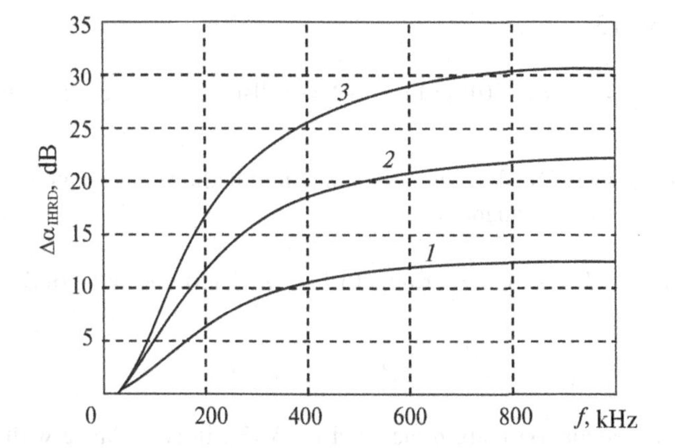

Fig. 5.17 Graphs of frequency dependence of line path attenuation increase due to IHRD on 110 kV overhead line conductors, by regions 1-I and II regions; 2-III region; 3-IV region

Stage 1 First, we define the parameters common to all adaptation steps used in calculations: attenuation of all elements of the HF path under consideration, except for attenuation of interphase wave $A(f_1)$ in the OPL which is defined according to the formula:

$$a_{\text{el.}} = a_{\text{line.end}} + a_{\text{LT}} + a_{\text{CD}} + a_{\text{HFcab}} + a_{\text{tap.line}} \quad (5.12)$$

In our example, we use typical values of components in (5.12) [12]: line end attenuation (modal conversion loss) $a_{\text{line end}} = 2.5$ dB; coupling loss caused by the line trap, $a_{\text{LT}} = 3$ dB; coupling loss caused by the coupling device, $a_{\text{CD}} = 1.5$ dB; attenuation of the HF cable $a_{\text{HFcable}} = 0.5$ dB; and attenuation caused by the tap line, $a_{\text{tap. line}} = 5.5$ dB.

Taking into account, these data: $a_{\text{el}} = 2.5 + 3 + 1.5 + 0.5 + 5.5 = 13$ dB.

Stage 2 Let us define overlapped attenuation of the equipment. It is defined for each adaptation step by the formula (5.1), considering K_{OPL} in (5.2). The calculation will be carried out taking into account that the noise level for the 110 kV overhead line is −38 dBm in 1 kHz band, nominal receiving band of the used DPLC equipment is 8 kHz, and corona factor *K*OPL which takes into account difference between white noise and corona noise is 2 dB (see Table 5.6).

Adaptation step 1:

$$p_{\text{RX.min1}} = (-38) + 10\,\lg(8) + 33 + 2 = +6\ \text{dBm}; a_{\text{overl.1}} = 35 - 6 = 29\ \text{dB}$$

Adaptation step 2:

$$p_{\text{RX.min2}} = (-38) + 10\lg(8) + 25 + 2 = -2\text{dBm}; a_{\text{overl.2}} = 35 + 2 = 37\text{dB}$$

Adaptation step 3:

$$p_{\text{RX.min3}} = (-38) + 10\lg(8) + 21 + 2 = -6\text{dBm}; a_{\text{overl.3}} = 35 + 6 = 41\text{dB}$$

Stage 3 Next, we proceed to calculating maximum allowable frequency at the first and subsequent steps of adaptation.

Adaptation step 1 Using the formula (5.13), we determine value of $A(f)$:

$$A(f_i) = p_{\text{RXmin.}i} - a_{\text{el.}} + a_{\text{marg.}i} \tag{5.13}$$

The value of attenuation margin included in (5.13), in accordance with the initial data, is assumed to be equal to $a_{\text{marg.1}} = 9$ dB. In this case, the value of $A(f)$ is independent of frequency (attenuation due to IHRD which causes this dependence is not considered) and is equal to:

$$A(f_1) = 29 - 13 - 9 = 7\text{dB}.$$

Using the formula (5.14) below, we determine attenuation per kilometer of the interphase wave corresponding to a certain value $A(f)$:

$$\alpha_{\Phi(f)} = \frac{A(f_1)}{L} = \frac{7}{75} = 0.93\,\text{dB}/\text{km}$$

Figure 5.16 shows the graphs of attenuation per kilometer for different ACSR conductors (outer diameter is 13.5 mm for ACSR-95, 15.4 mm for ACSR-120, 18.8 mm for ACSR-185, and 21.6 mm for ACSR-240). For ACSR-120 conductor, the attenuation per kilometer is $\alpha_{\Phi(f)} = 0.093$ dB/km at a frequency of 280 kHz. This frequency is a maximum allowed frequency for operation at the first step $f_{\text{max1}} = 280$ kHz.

Adaptation step 2 To make the calculation for this step, we will use the values already defined above:

- Attenuation of all elements of the HF path under consideration, except for attenuation of interphase wave in the OPL a_{el} = 13 dB
- Overlapped attenuation of the equipment $a_{overl.\,2}$ = 37 dB

Then the iterative process begins.

The first iteration We use the expert estimate of the maximum allowable frequency first approximation equal to f_1 = 280 kHz. For this frequency, Δa_{IHRD} defined according to Fig. 5.17 is equal to Δa_{IHRD} = 16 dB In accordance with (5.6), a_{marg2} = 16 dB.

Taking this into account, $A(f)$ is equal to:

$$A(f_2) = 37 - 13 - 16 = 8\text{dB}.$$

Using the formula (5.14), we define attenuation per kilometer of the interphase wave corresponding to a certain value $A(f)$:

$$\alpha_{\Phi(f)} = \frac{A(f_2)}{L} = \frac{8}{75} = 0.107\text{dB / km}$$

According to Fig. 5.16, $\alpha_{\Phi(f)}$ = 0.107 dB/km corresponds to the frequency f_2 = 340 kHz. This is higher than the frequency used as the first approximation, and it is necessary to select the second frequency approximation and to run the second iteration. Here, we will not do this because maximum allowed frequency at the second step is, in any case, more than at the first step and cannot be used as maximum allowed frequency of the channel $f_{max.chan}$.

Adaptation step 3 Since the overlapped attenuation of the channel at the third adaptation step is more than at the second step (41 instead of 37 dB), maximum allowable frequency of the third step will be even more than for the second step and will not have effect on resulting maximum allowable frequency of the channel. So we will not calculate this parameter.

Table 5.8 Results of calculating parameters of the PLC channel with bit rate adaptation

Parameter	Bit rate, B_{DPLC}, kbps/step		
	48/1	36/2	24/3
SNR for AWGN, dB	33	25	21
Minimum reception level $p_{RX.min}$, dBm	6	−2	−6
$A(f)$, dB	7		
a_{el}, dB	13		
$a_{overl.}$, dB	29	37	41
$a_{marg.req.}$, dB	9	16	16
a_{marg}, dB	9	17	21
$f_{max.chan}$ = 280 kHz; K_{OPL} = 2 dB; p_{TX} = 35 dBm; Δp_{noise} = 9 dB; a_{IHRD} = 16 dB			

Thus, $f_{max1} < f_{max2} < f_{max3}$ and maximum allowed channel frequency is assumed to be $f_{max.chan} = f_{max1} = 280$ kHz.

It should be noted that obtained ratio between f_{max1}, f_{max2}, and f_{max3} is largely determined by the initial data for the overhead line designing. So, for example, if the phase conductor of the overhead line is ACSR-185, but not ACSR-120, then maximum allowable frequency of the second adaptation step in this path would be the lowest one.

Step 4 Final calculation of the channel parameters

In the final calculation of the channel, we are interested in attenuation margin which the channel has when working at different steps of adaptation.

Let us assume for simplicity that the upper edge frequency of the nominal reception frequency band is equal to maximum allowed frequency of the channel.

It is necessary to define the margin for this case when the channel is running, for each adaptation step. To do this, first, for each step, we define estimated attenuation of the HF path at a frequency of 280 kHz and overlapped attenuation of the equipment.

According to the data obtained above when determining maximum allowable frequency, these parameters are equal to $a_{HFpath.} = 13 + 7 = 20$ dB; $a_{overl.1} = 29$ dB; $a_{overl.2} = 37$ dB; $a_{overl.3} = 41$ dB.

Using the formula (5.8), we get attenuation margin for each step:

- For the first step: $a_{marg1} = 29 - 20 = 9$ dB
- For the second step: $a_{marg2} = 37 - 20 = 17$ dB
- For the third step: $a_{marg3} = 41 - 20 = 21$ dB

The results obtained show that each lower step extends the operability limits in comparison with the previous step by the value of the difference SNR_{min}.allow for these steps. This also follows from the formula (5.10). It should be also noted that it is not possible to separate the resulting attenuation margins into components Δp_{noise} and Δa_{IHRD}. This is possible only when defining the required margin minimum allowable value using (5.6) or (5.7). The calculation results are summarized in Table 5.8.

When using an accurate method for HF path attenuation calculation, the following circumstances should be considered to define maximum allowable frequency.

The first thing to take into account is that the calculated noise level for simplified calculation methods is defined for the point of the OPL conductors connection to its ends. In this case, overlapped attenuation of the equipment, with which estimated value of the HF path attenuation is compared when defining maximum allowable frequency of the channel, is defined without attenuation caused by line traps and coupling devices (LTCD) on the receiving end of the OPL. This is because these devices contribute the same attenuation to both the HF signal receiving path and noise receiving path, and not taking them into account does not affect the signal to noise ratio at the receiver input.

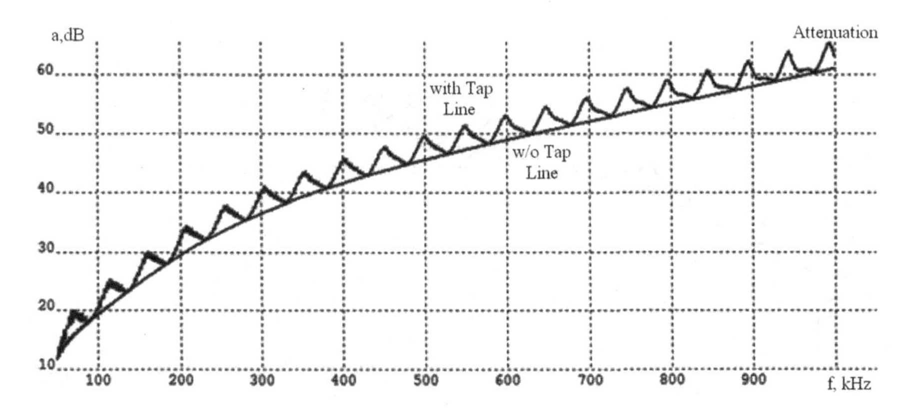

Fig. 5.18 Frequency response of HF path with IHRD, with LTCD, with tap line, and without tap line

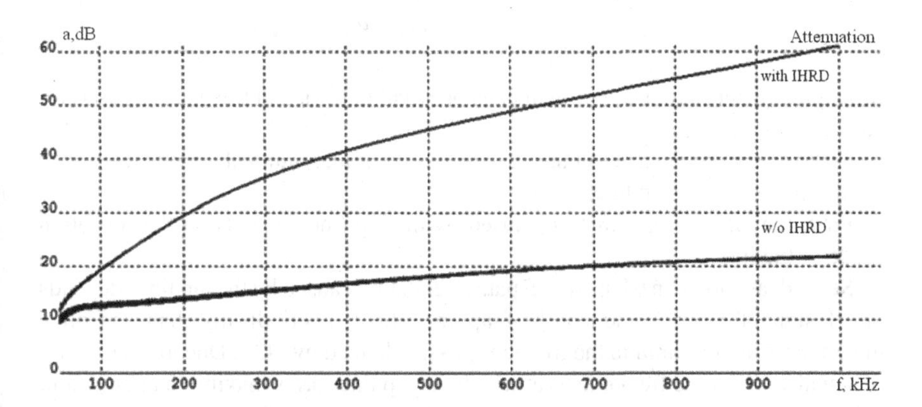

Fig. 5.19 Frequency response of HF path with LTCD and without tap line, with IHRD, and without IHRD

Now let us make a similar calculation $f_{\text{max.chan}}$ but using accurate calculation methods. Annex 1 contains description of the WinTrakt calculation model of the OPL under consideration.

At the same time, attenuation of the HF path calculated using WinTrakt software considers line traps and coupling devices at both ends of the HF path. Therefore, at comparison of overlapped attenuation with the calculated attenuation of the HF path, it is necessary whether to previously increase overlapped attenuation on the value of attenuation introduced by LTCD at one side of the HF path or to decrease calculated attenuation of the HF path by the same amount.

To define attenuation of the LTCD at one end of the path $a_{\text{WT(LTCD)}}$ using WinTrakt software, calculation is performed considering all LTCD devices installed ($a_{\text{WT(path with LTCD)}}$) and not taking into account the LTCD in the coupling phase ($a_{\text{WT(path w/o LTCD)}}$). Based on the results of these calculations, attenuation of the LTCD devices ($a_{\text{WT(LTCD)}}$) is calculated for one end of the HF path:

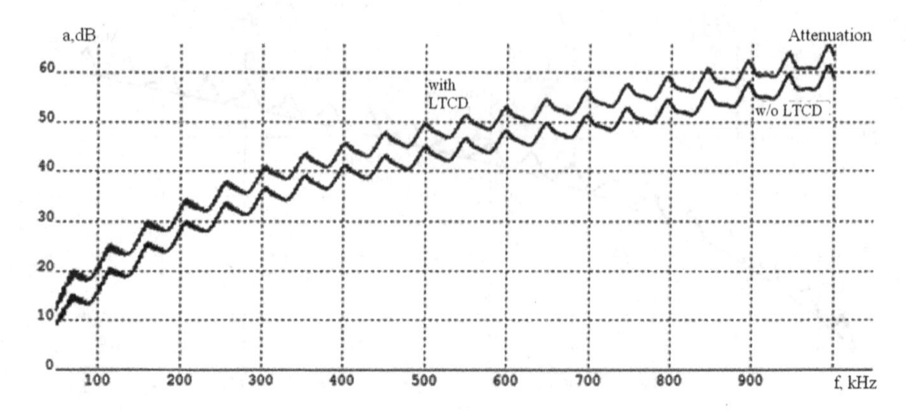

Fig. 5.20 Frequency response of HF path with tap line, with IHRD, with LTCD, and without LTCD

$$a_{\mathrm{WT(LTCD)}} = 0.5\left(a_{\mathrm{WT(HFpath\ with\ LTCD)}} - a_{\mathrm{WT(HFpath w/oLTCD)}}\right)$$

(doing this is most convenient by means of Excel tables), and this value is used for calculations.

The second circumstance concerns calculation of maximum allowable frequency for the HF path with tap line.

First, the HF path attenuation caused by the tap line depends on the operation mode of the line.

Second, as shown in Chap. 4, HF path attenuation caused by the tap lines depends on HF signal frequency and length of tap line ratio and periodically changes from a maximum to a minimum in the frequency range defined by (4.8). Due to uncertainty of actual value of the frequencies at which the tap line forms maximum attenuation, it is acceptable to assume that, all frequencies within the frequency range under consideration, attenuation caused by the tap line is equal to its maximum possible value.

Taking this into account, calculating maximum allowable channel frequency using accurate calculation methods becomes somewhat more complicated and actually becomes a combination of accurate and simplified calculation methods. This combination requires definition of all HF path element attenuation using WinTrakt software while considering specific design of the OPL, coupling device, CC, HF cable, and LT. At that, calculation using accurate methods should include the following tasks:

Task 1 Evaluation of full HF path attenuation using WinTrakt software, considering ice coating ($a_{\mathrm{WT(path\ with\ IHRD)}}$).

Task 2 Evaluation of HF path attenuation using WinTrakt software according to Task 1 but without LTCD at the coupling phases at the ends of the HF path ($a_{\mathrm{WT(path\ with\ IHRD\ w/o\ LTCD)}}$).

Task 3 Evaluation of the HF path attenuation using WinTrakt software according to Task 1 with the exception from the HF path of tap line (but with LTCD at the ends of OPL) ($a_{\text{WT(path with LTCD, IHRD w/o tap line)}}$). Graphs of HF path attenuation considering IHRD, both with and without a tap line, are shown in Fig. 5.18.

Task 4 Evaluation of the HF path attenuation using WinTrakt software according to Task 3, except of IHRD effect ($a_{\text{WT(with LTCD, w/o IHRD, tap line)}}$). Graphs of frequency response for the path without tap line, and also without IHRD and with IHRD, are shown in Fig. 5.19.

Task 5 Evaluation using an Excel table (or another method) of the LTCD attenuation at one end based on entered into Excel table calculation results according to Tasks 1 and 2 ($a_{\text{WT(LTCD)}}$). Graphs of frequency response for the path with IHRD and tap line with LTCD and without LTCD are shown in Fig. 5.20.

Task 6 Evaluation using an Excel table of maximum attenuation caused by the tap line based on the calculation results entered into Excel table (or by another method) according to Tasks 1 and 3.

Task 7 Evaluation using an Excel table (or another method) of maximum allowed channel frequency as the frequency for which conditions are met, as a result, for the first adaptation step:

$$a_{\text{overl.1}} - a_{\text{tap.line}} + a_{\text{LTCD}} - a_{\text{marg.req.1}} - a_{\text{HFpath w/o tap.line w/o IHRD}} \approx 0 \tag{5.15}$$

- For all subsequent adaptation steps

$$a_{\text{overl.}i} - a_{\text{tap.line}} + a_{\text{LTCD}} - a_{\text{marg.req.}i} - a_{\text{HFpath w/o tap.line with IHRD}} \approx 0 \tag{5.16}$$

At that, in accordance with the task, $a_{\text{marg.req1}} = 9$ dB and $a_{\text{marg.req2}}$ and $a_{\text{marg.req3}} = \Delta a_{\text{IHRD}}$, but not less than 9 dB.

As a result of calculations performed in accordance with the steps described above, we get: $a_{\text{tap.line}} = 3.5$ dB; $a_{\text{LTCD}} = 2.25$ dB.

At that, while keeping in mind that $a_{\text{overl.1}} = 29$ dB; $a_{\text{overl.2}} = 37$ dB; $a_{\text{overl.3}} = 41$ dB, equality (5.15) for the first adaptation step and equality (5.16) for the second and third adaptation steps are satisfied at frequencies $f_{\max 1} = 540$ kHz, $f_{\max 2} = 320$ kHz and $f_{\max 3} = 360$ kHz. These frequencies are the maximum allowed frequencies for the 1-st, 2-nd and 3-rd steps of adaptation, respectively.

Thus, maximum allowed channel frequency is assumed to be 320 kHz.

Attenuation due to IHRD is defined for the resulting frequency of 320 kHz (as the difference of calculation results according to tasks 3 and 4 for this frequency). This attenuation is 21.9 dB.

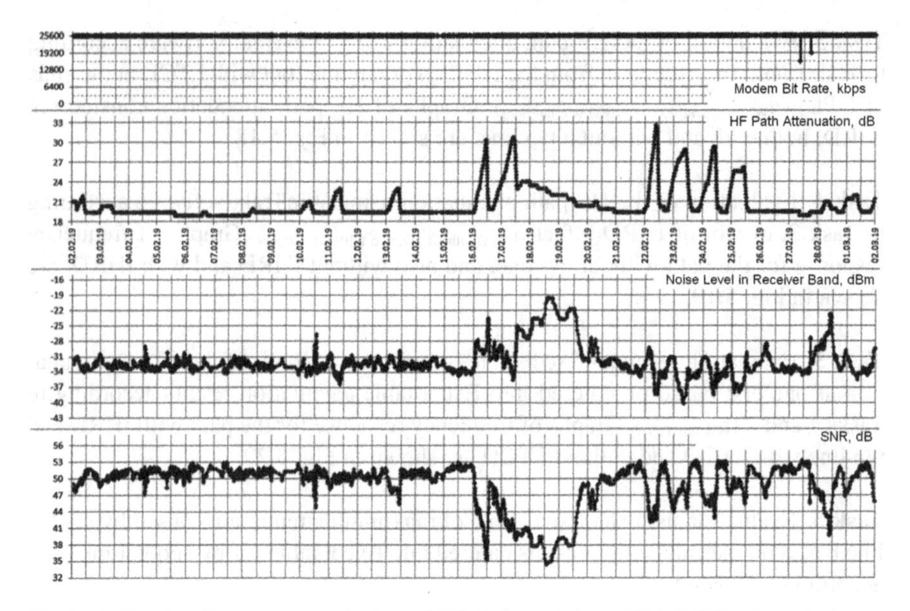

Fig. 5.21 Graphs of long-term monitoring of DPLC channel along 110 kV OPL. Example 1

Let us define the attenuation margin at each adaptation step. Using the formula (5.10), it is possible to evaluate attenuation margins for each channel adaptation step. They are equal to:

- For the first step: $a_{\text{marg}1} = 13.94$ dB
- For the second step: $a_{\text{marg}2} = 21.9$ dB
- For the third step: $a_{\text{marg}3} = 25.9$ dB

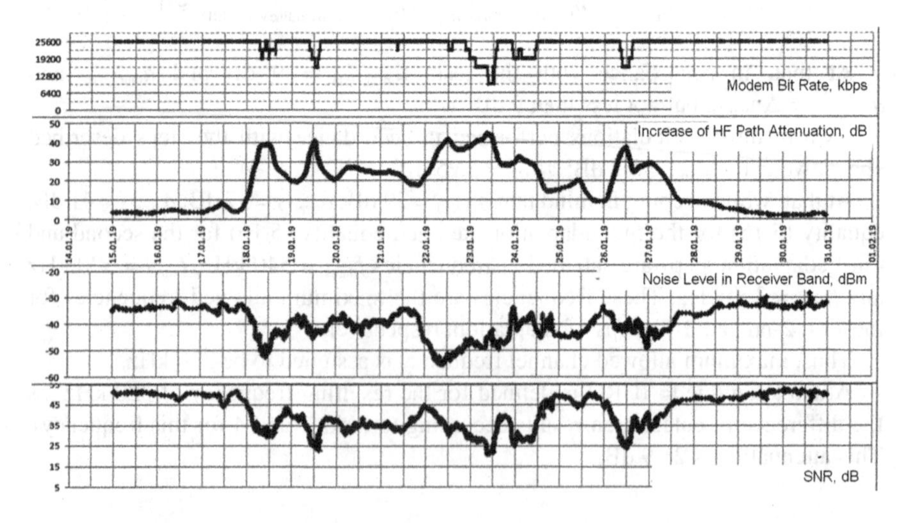

Fig. 5.22 Graphs of long-term monitoring of DPLC channel along 110 kV OPL. Example 2

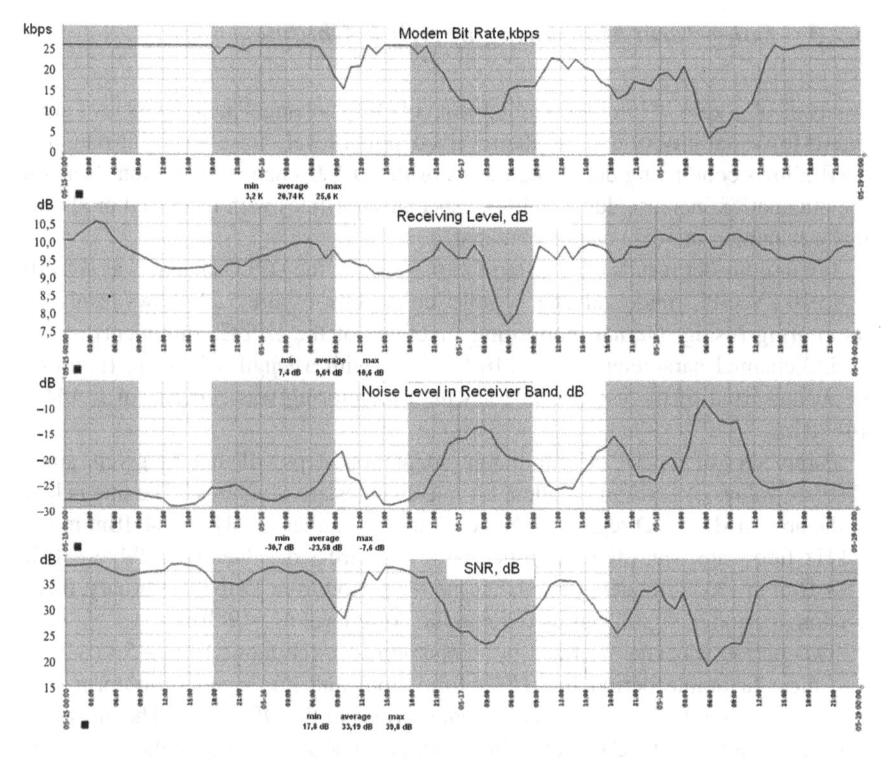

Fig. 5.23 Graphs monitoring of DPLC channel along 220 kV OPL

Significant difference of the results of maximum allowable channel frequency calculations using simplified and accurate calculation methods is due to the fact that, when using accurate methods, attenuations of all elements of the HF path, in particular, the LT, CC, CD, HF cable, and tap line, are clarified.

Correctly designing DPLC channels is the key to their reliable operation. However, all calculation methods are based on statistics of occurrence of certain events which affect the HF path parameters. Hardly ever, the calculated results have a significant deviation from the actual values. This is mainly due to incomplete data on operating conditions and abnormal weather changes during the year, when the actual parameters of the HF path are significantly worse than the calculated ones, for example, when IHRD fall on a section or sections of power lines exceeding in total calculated value of 30 km.

In the most difficult cases, if the DPLC channel operation results are not satisfactory, results of long-term attenuation increase monitoring should be used as input data for project correction. Project correction should be based on existing monitoring results with possible shift of the operating frequency to an area with less HF path attenuation of or with bit rate reduction, with change of the HF path structure, replacement of HF bridges with additional DPLC channels, etc. This approach will significantly improve design quality and efficiency of communication channels.

5.7.3 Long-Term Monitoring of DPLC Channels

Analysis of graphs of long-term monitoring of DPLC channel parameters is of great interest. Availability of data on the received signal level, noise level, bit rate, and SNR allows conducting detailed analysis of the DPLC channel operation. Data on the attenuation increase due to IHRD, especially for lines with increased or abnormal attenuation increase can be obtained.

Let us consider results of long-term monitoring of the DPLC channels along 110 and 220 kV OPLs under adverse weather conditions. Figure 5.21 shows results of monitoring (using internal monitoring functions of the DPLC equipment) of the DPLC channel parameters along 110 kV OPL with a length of 74 km (nominal frequency band of the equipment for which monitoring was carried out is 144... 148 kHz).

Parameters of the DPLC equipment adaptation steps will not be given, since absolute values of SNR and modem bit rate changes are of interest. The line is laid in the area of III IRHD region. Noise level at dry weather is −31... −34 dBm in the 4 kHz frequency band that complies with standardized values (see Table 4.10): −38 dBm in 1 kHz frequency band. During the period from 17 to 20 February, there is a long period of significant increase in the noise level to −19 dBm.

This period was characterized by temperatures in the range from −5 to +2 °C (the most favorable period for IHRD formation) and prolonged snowstorm (the cause of increased noise level) in different sections of the OPL route. The most difficult conditions for PLC channels in winter are IHRD formed near the substation where transmitter is located and heavy precipitation (snowstorm, rain) around the substation where receiver is located. These conditions are especially probable for long lines. In this case, the communication channel remained operable with the primary bit rate without reducing it, except for two cases of short-term bit rate reduction for unknown reasons.

Figure 5.22 shows results of the DPLC channel operation monitoring at another 110 kV OPL. The line is laid in the area where IHRDs in the form of rime ice are often observed; we recall that this type of IHRD makes the largest contribution into attenuation (see Fig. 4.15). Also, during the observation period, there was heavy snow several times. It can be seen at the graphs shown in Fig. 5.22 that the attenuation increase is more than 40 dB in several time periods. It is clearly seen that, when the attenuation increases due to IHRD, the level of noise decreases, so that SNR value changes in the range of not more than 30 dB. The graph of bit rate change shows that, at high SNR, data bit rate is 25.6 kbps. The lowest bit rate at which the modem worked was 9.6 kbps.

Figure 5.23 shows graphs of the PLC channel monitoring along 220 kV OPL with a length of 210 km during heavy snowfall near the substation (nominal frequency band in which monitoring was carried out is 196... 200 kHz). It can be seen from upper graph of the bit rate how the equipment adapts by transfer from the primary bit rate of 25.6 kbps to 3.2 kbps during heavy snowfall. Period of significant bit rate reduction begins around 3 a.m. and continues until 12 p.m. At that, there is a strong increase of the noise level from −20 dBm to −8 dBm with a slight decrease of reception level; finally, due to influence of both factors, SNR decrease was 14 dB: from 32 dB to 18 dB.

In the given example, the equipment configuration provides transmission of three voice signals and a telecontrol data at the maximum bit rate.

As the bit rate decreases, voice transmission is disabled according to priorities. For a period of about 2 hours, voice channels were disabled, and only telecontrol data were transmitted at a rate of 3.2 kbps.

Measurements for 220... 330 kV OPL show that for a properly designed channel, absolute SNR value can vary in the range from 50 to 18 dB. At that, high SNR value is 34... 50 dB, low SNR value is 18... 26 dB; this is caused by various destabilizing factors, for example, the HF path in attenuation increase due to IHRD: 9... 18 dB, effect of rain, thunderstorms, wet snow, etc.

References

1. IEC 60870-5-101 Telecontrol equipment and systems – part 5–101: transmission protocols – companion standard for basic telecontrol tasks, International standard, IEC (2003)
2. IEC 60870-5-104 Telecontrol equipment and systems – part 5–104: transmission protocols – network access for IEC 60870-5-101 using standard transport profiles International standard, IEC (2006)
3. Merkulov, A.G., Shuvalov, V.P.: On the issue of IP header compression application in high voltage digital power line carrier channels. Paper presented at the 2016 IEEE International Siberian Conference on Control and Communications (SIBCON), Moscow, Russia, 12–14 May 2016, pp. 1–5. https://doi.org/10.1109/SIBCON.2016.7491656
4. Merkulov, A.G., Shuvalov, V.P.: The method of narrow band PLC channels throughput increase. Paper presented at the IEEE International Energy Conference. Limassol, Cyprus, 3–7 June 2018
5. Fortunato, E., Garibbo, A., Petrolino, L.: An experimental system for digital power line communications over high voltage electric power lines – field trials and obtained results. Paper presented at the 7th ISPLC Symposium, Japan, Kyoto 2003, 26–28 March 2003
6. Merkulov, A.G.: Technologies of packet networks organization via high voltage digital power line carrier channels (in Russian). Hot Line Telecom, Moscow (2017)
7. Merkulov, A.G., Shuvalov, V.P.: Analytical study of principles organization of convergent packet DPLC networks with transition from frame relay to IP technology with examples of projects implemented in Kazakhstan. Paper presented at the IEEE International Siberian Conference on Control and Communications, Omsk, Russia, 21–23 May 2015
8. Raviola, A., Garoti, G.: Reliable architecture for power system operational communications integration of digital PLC & ATM. Paper presented at the CIGRE D2 Colloquim B03, Rio de Janeiro, Brasil, 2003
9. Baker, F., Polk J., Dolly M.: A differentiated services code point (DSCP) for capacity-admitted traffic (updates RFC 4542 and RFC 4594), RFC 5865, IETF (2010)
10. Collins, D.: Carrier Grade for Voice over IP. McGraw Hill Networking, New York (2004)
11. Shkarin, Y.P.: Measurements in high voltage PLC communications. WinTrakt and WinNoise software for computation of high frequency characteristics of the high voltage power lines and corona noise (in Russian). Analytic, Moscow. http://www.analytic.ru/articles/art450.pdf (2015). Accessed 01 June 2020

12. Standard of Organization. Guidelines for selection of frequencies of high-frequency channels along 35, 110, 220, 330, 500 and 750 kV electric power transmission lines. Federal Grid Company of the Unified Energy System of Russia, Moscow (2010)
13. Braude, L.I., Philippov, A.A., Kharlamov, V.A., Shkarin, Y.P.: Research of discrete information transmission along power line carrier channels (in Russian), Electr. J. (8), Moscow, 8–14 (1999)
14. Mujčić A., Suljanović, N., Zajc, M., Tasič, J.F.: Error probability of MQAM signals in PLC channel. Paper presented at the 11th International Electrotechnical and Computer Science Conference ERK 2003, Portorož, Slovenia, 25–26 September 2003

Annex No. 1. General Description of WinTrakt and WinNoise HF Path Calculating Programs

Features of WinTrakt and WinNoise programs.

WinTrakt allows calculating the parameters of a HF path in the frequency range from 10 to 1000 kHz. These parameters provide a full description of a HF path:

- HF path and linear path attenuation
- Input impedance (for both sides of a HF path)
- Return loss (for both sides of a HF path)
- Group time delay

Parameters are calculated in a cycle with a set frequency.

The software allows to:

- Create a HF path scheme of any complexity level and calculate path parameters for any coupling scheme to phase conductors and isolated groundwire of OPL and to phase conductors and sheaths of CPL
- Consider the effect of ice coating at any section of any OPL belonging to HF path
- Consider periodic OPL non-homogeneities due to the effect of metal and reinforced-concrete towers of the OPLs and groundwires during their grounding on each tower
- If necessary, calculate modal parameters of any power line (both overhead and cable) belonging to HF path scheme

There is a database for the calculation model of a HF path. It is completed by a user as calculations are conducted and can be applied for calculations of other paths.

Calculation results are presented as tables and/or graphs which can be inserted from Windows clipboard to any editor if needed. On one figure, one can present the results of several calculations both for one and several HF paths.

WinNoise allows calculating parameters of corona noise at alternating and direct current OPL with the voltage of 110 kV and above. The calculation operates in the frequency range from 10 to 1000 kHz. These parameters provide a full description of corona noise:

A. G. Merkulov et al., *High Voltage Digital Power Line Carrier Channels*,
https://doi.org/10.1007/978-3-030-58365-1

Average root mean square noise levels in HF paths with any schemes of coupling to phases and groundwires of overhead power lines under different conditions impacting the noise level along the line path (e.g., air pollution, weather conditions, or altitude). Parameters can be calculated in a frequency cycle.

Dependence of relative root mean square noise voltage on phase of industrial frequency voltage (only for alternating current lines). Such calculations are important for DPLC channels.

PLC channel.

The software can factor in the following into calculations:

Generation function defined using the method approved in Russia or recommended by CIGRE, considering the features of generation function calculation applying these methods. Noise generation function calculation method recommended by CIGRE is differentiated for alternating current OPL and direct current transmission (DCT). At that, the method assumes that during DCT under adverse weather conditions the noise level does not increase, but decrease, in opposite to alternating current OPL.

Change of generation function on OPL conductors for certain circumstances occurring at different line path sections. By setting various generation function values at different line path sections one can, for instance, calculate corona noise level for lines installed in the mountains or along a highway with local air pollution near industrial facilities. In addition, precipitation over a part of the line may increase generation function values.

WinNoise, the same way as WinTrakt, offers creation of a database (which can be common for both programs): the identical way of calculation result presentation and pasting to any editor from Windows clipboard.

Software calculation results were extensively tested and compared to actual paths. They reveal high coincidence level with the experiment outcomes.

Software algorithms.

Algorithms of both programs are based on a modal theory of signal propagation along heterogeneous multiconductor power lines considering the effect of earth (earth presentation models can be heterogeneous and half-infinite or two-layer).

When describing the noise sources on OPL conductor, WinNoise uses experimental data on the normalized pulse spectrum of streamer corona and generation function characterizing the source intensity.

Let us consider the issues related to creation of HF path calculation model, conducting of calculations and presentation of calculation results using WinTrakt program as an example (as they are the same for WinNoise program).

HF path calculation model is built using a graphical menu incorporating the following elements:

1. TL – total end line load. Impedance at the end of power line during calculation of linear path attenuation of the power line itself disregarding the attenuation caused by line trap and coupling devices.
2. CD – coupling devices. Element describing the parameters of coupling devices (line matching units) including coupling capacitor in scheme and high-frequency cable between the PLC equipment and CD.

3. OL – overhead power line. The element includes the description of geometric parameters of OPL towers, types, and features of phase conductors and groundwires, specific soil resistivity, and description of ice, hoarfrost, and rime deposits (IHRD).
4. CL – cable power line. The element describes cable line parameters.
5. Tr –transposition.
6. ShA – shunt admittance. The element is used to describe line traps at the ends of HF path and OPL damages related to OPL conductors short circuits to earth and between wires.
7. SI – series impedance. The element is used to describe line traps included into OPL conductors in series (at any site, e.g., in a tap line) and OPL damages related to a phase conductor breakage.
8. ET – load at the end of tap line. The element is used to describe the parameters of a branch substation connected at the end of a tap line. It factors in substation equivalent capacitance and OPL operation mode: connected, disconnected, and grounded at a substation. The element appears when the tap line scheme description is finished.
9. HFB – HF bridge (of an intermediate substation, cable line or field HF bridge). This element includes the description of HF bridge scheme and an intermediate substation.
10. Tap – tap line. The element appears at the beginning of tap line description.
11. CNC – the number of conductors change. The element indicates the beginning of a section with the changed number of conductors, e.g., when a groundwire hang at a certain length of the main path of the OPL line or other OPL passes.
12. CND – the number of conductors duplication. This element is commonly used to describe a HF path with a HF bridge transmission intermediate substation when both lines in a path approaching this SS form a constant bearing or hang on common double-circuit towers.
13. EA – end of connected circuit. The element entitled End of connected circuit is used either to indicate the end of describing the connected circuit of "additional" conductors at the beginning of the section with the changed number of conductors or as the last element on a section with duplicated number of conductors.
14. – beginning of description of OPL branching. This type of non-homogeneity allows to make a circuit branch on a multicircuit OPL path (e.g., double-circuit) at any multicircuit OPL section into two independent lines with different schemes and then unite these lines into a multicircuit OPL. An example is a double-circuit OPL where at a certain part of OPL circuits break up into separate paths of various length.

15. – end of a OPL branching section.

Apart from these designations, the following control buttons are used in the user interface:

1. – warnings shown during HF path model formation.
2. – database. Database includes a library of different HF path elements: CD, HF cables, line traps, OPL towers, types of phase conductors, groundwires, and underground cables. A user may complete the database independently by forming a certain element according to its actual technical parameters.
3. – intermediate editing results saving.
4. – successful end of editing.
5. – end of editing without saving the results.

Now let us consider creating a calculation model of a HF path in WinTrakt software. OPL scheme is provided in Chap. 5, Fig. 5.14.

A HF path model is presented graphically. Figure A1.1 presents a final view of the created model of HF path in question. It is rendered by sequential input of the necessary elements from the menu with the set of HF path elements described above (Digits below element abbreviation indicate the element input sequential number e.g., 3 of element). After entering the next element, input data on this element is also entered.

In our example, the first HF path element is a coupling device . Figure A1.2 shows a window for entering input data for element . One shall enter the type of coupling device (line matching unit), HF cable, and length of HF cable, and its impedance for the phase B the CD is connected to in the scheme in question. Lines for phases A and C not connected to the CD are left blank. On the right, you can see a window for setting the input data for the elements of CD unit scheme (bandwidth 48–1000 kHz; coupling capacitor (C1) of 6400 pF).

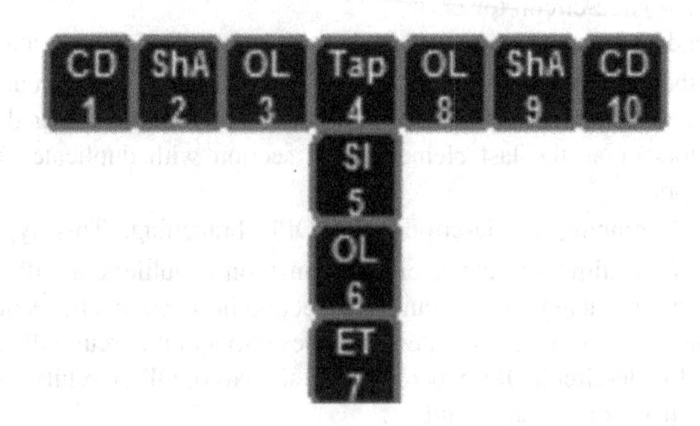

Fig. A1.1 Calculation model of a HF path with a tap line

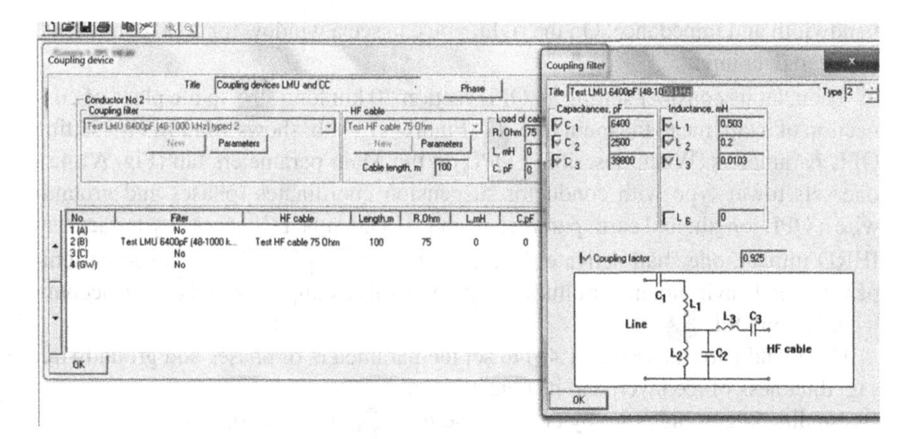

Fig. A1.2 Description of Coupling device elements

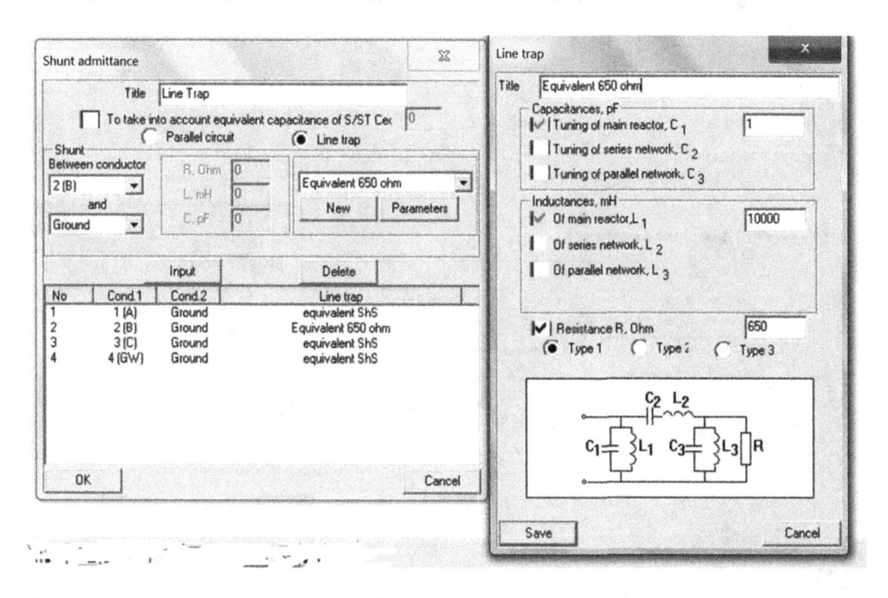

Fig. A1.3 Description of a Shunt admittance element representing a line traps on the left end of HF path

The next HF path element is Shunt admittance . Figure A1.3 shows a window for entering input data for element providing line traps on SS 1 and input impedance of this SS. In the calculation, SS input impedance is zero (it is a large SS with a many 110 kV OPLs). In view of the above, an “Equivalent 650 ohm” line trap is introduced between phase B and the earth, with impedance of 650 ohm throughout the entire PLC frequency range. In all the remaining phases, a “ShS equivalent” with 0.1 ohm impedance is installed. If exact line trap parameters are known, instead of a 650 ohm equivalent, one shall use a line trap with compatible blocking bandwidth and impedance. On the right, you can see a window for "650 equivalent" line trap data input.

Then, let us consider setting an OPL section 30 km long to the place of connection of a tap line to the main line. Figure A1.4a, b, shows windows for setting OPL parameters. When describing OPL on the Main parameters tab (Fig. A1.4a), one sets tower type with conductor suspension coordinates (phases and groundwire); OPL length; and earth parameters and IHRD impact (if necessary to factor in IHRD impact, one shall activate the relevant field and set the temperature and environmental pollution factor). In this example, one takes into account ice layer on OPL.

Use Conductor tab (Fig. A1.4b) to set the parameters of phases and groundwire, sag, thickness of ice layer, and its density in cm).

Tap line beginning is set by entering element (Fig. A1.5).

Then, tap line description begins. The beginning is initiated by entering a Series impedance element to describe a line trap installed in Phase B (Fig. A1.6).

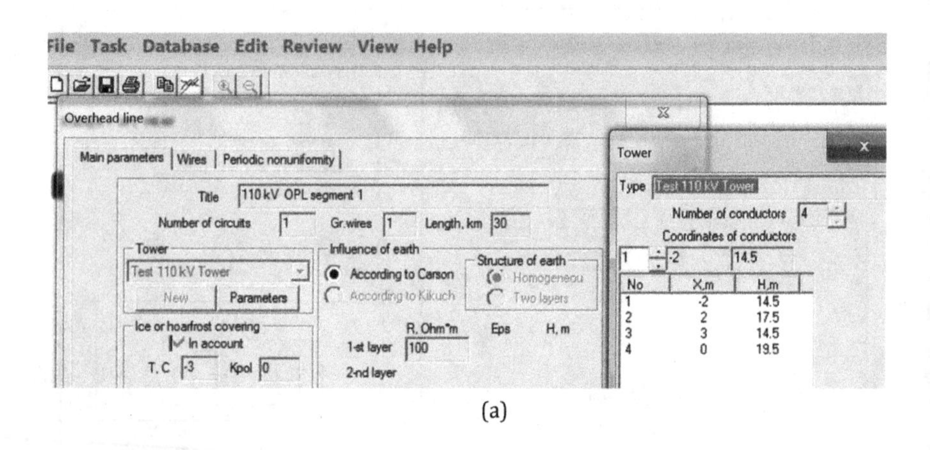

(a)

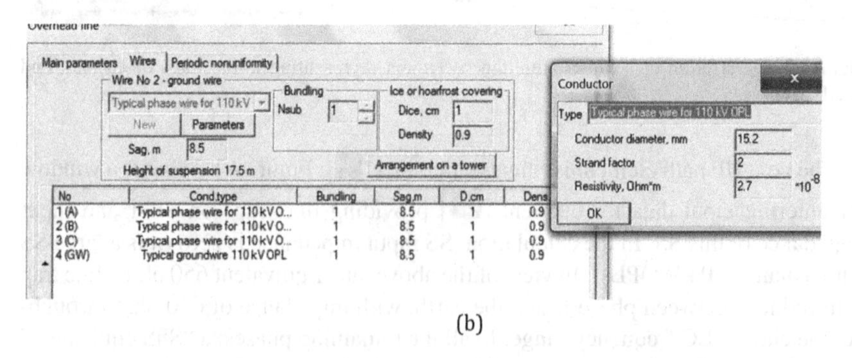

(b)

Fig. A1.4 (**a**) Description of OPL element from the path beginning to the place of tap line connection: main parameters; (**b**) Description of OPL elements conductors

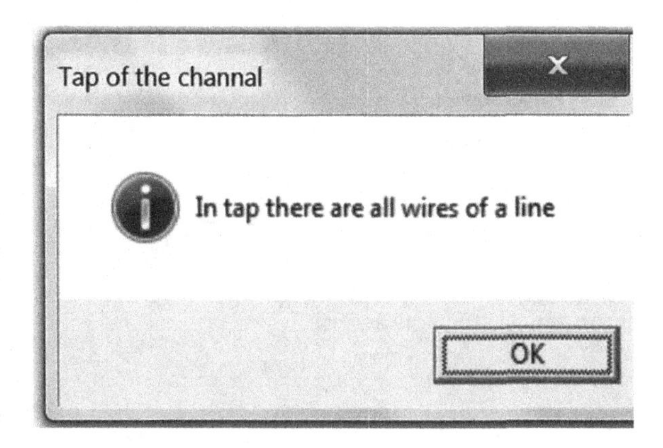

Fig. A1.5 Element of description tap line

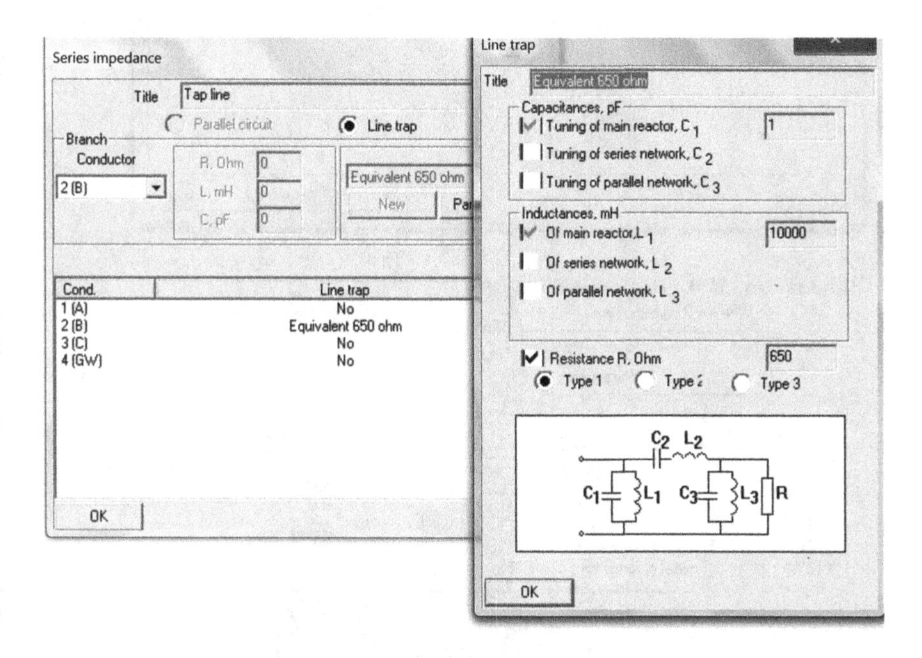

Fig. A1.6 Series impedance element description

The next step is describing an OPL section from the point of line trap suspension to the branch substation at the end of the tap line OL 6 (Fig. A1.7a, b). According to the example, the type of phase conductors and towers over the entire OPL length is the same (generally, WinTrakt allows describing OPL sections with different conductors and towers setting the relevant line elements OL 2 OL 3 OL 4 in series). Tap line used in the example is 3 km long. IHRD is not taken into account for this and subsequent OPLs as the IHRD on the necessary OPL section length in 30 km was considered during previous OPL section description before to the tap line.

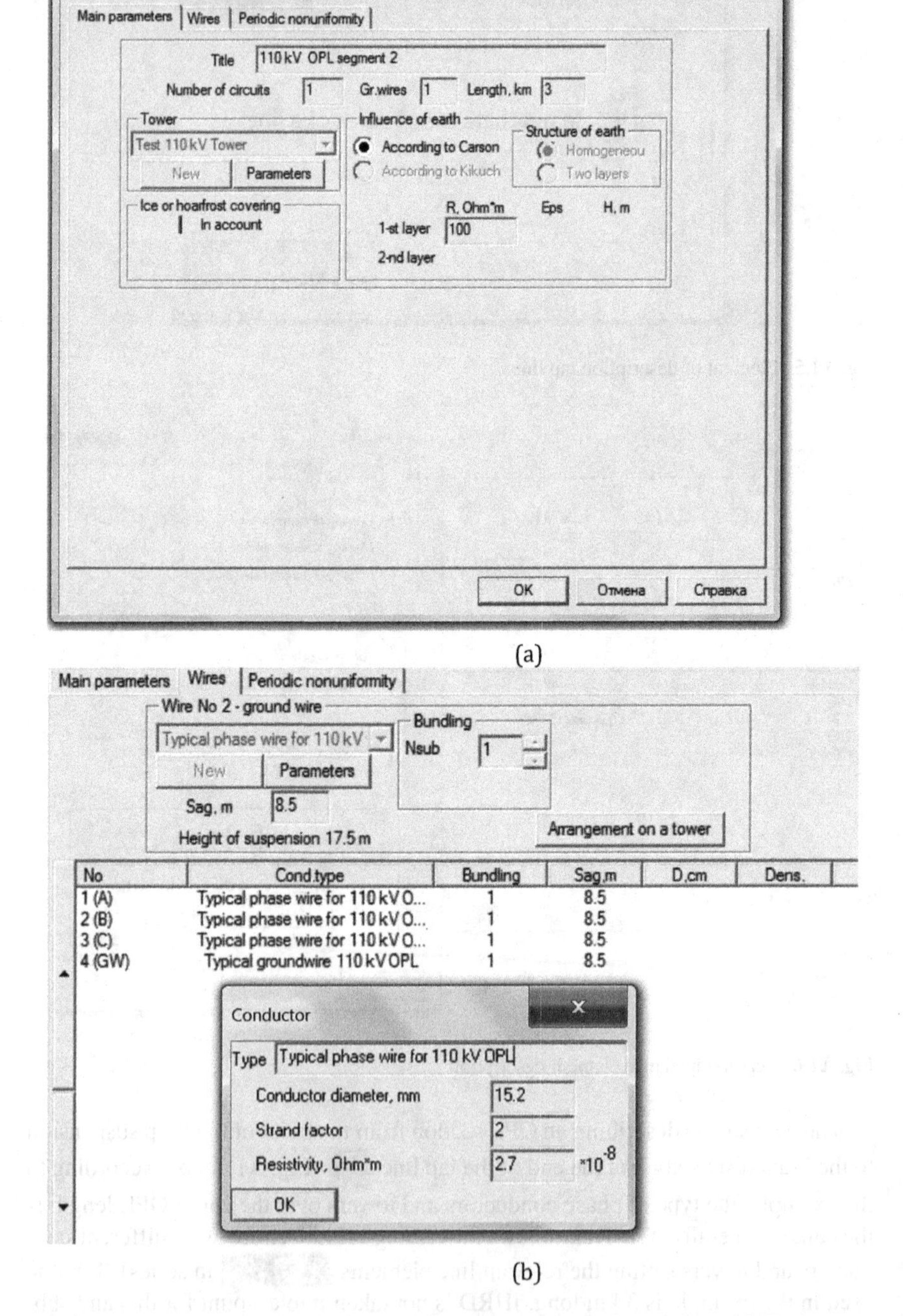

Fig. A1.7 (**a**) OPL tap line description: Main parameters tab. (**b**) Description of OPL elements: conductors

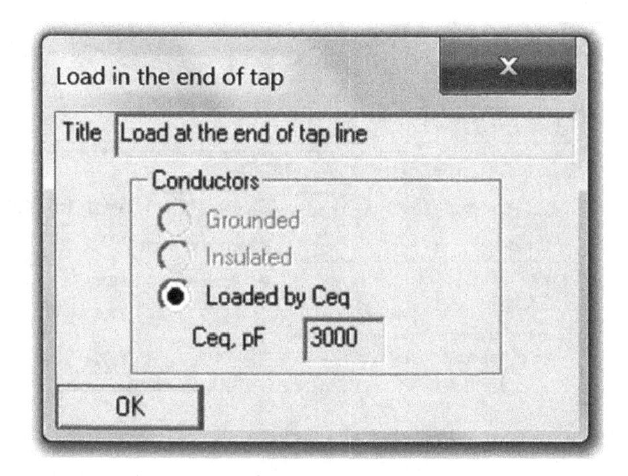

Fig. A1.8 Element of the end of the tap line describing line load

The end of tap line describing the branch substation input impedance (load at the end of the branch substation) is described by introducing ET 7 element (Fig. A1.8). In our case, the line equivalent capacitance of 3000 pF is applied as a recommended load.

After the tap line description, let us proceed to the description of the rest of HF path part from the tap line connection to the right end of HF path. Let us begin from an OPL OL 8 section 45 km long (Fig. A1.9). IHRD is not taken into account in this OPL, the same as in OL 6.

One needs to describe data input for elements ShA 9 and CD 10, after which HF path calculation model (Fig. A1.1) configuration will be completed. In our example, let us assume that line traps and coupling devices at both ends of HF path are the same. Hence, elements ShA 9 and CD 10 are completed in the same way as elements CD 1 and ShA 2. The difference can be found only in the input data volume when entering CD 1 and CD 10. In the element at the beginning of the path (CD 1), one shall additionally complete the set names of phases A, B, and C (Phasing button).

When the path is completely formed, press END to save the model into a separate file. It can be used to calculated frequency dependencies of HF path parameters. Figure A1.10 shows a window with settings of calculation of frequency dependencies of HF path parameters. In our example, it is a coupling scheme of phase B-to-earth with a calculation in the range of 50–1000 kHz and 1 kHz step. Figure A1.11 shows a window with the selection of calculation results and form of review graphs or tables. The software allows forming frequency response graphs, defining group time delay, input impedance, and return loss. Data can be presented in a table either graphs. Calculation results can be found on Fig. A1.12 (attenuation), Fig. A1.13 (input impedance on the left side, $Z_{(\mathrm{left})}$), and Fig. A1.14 (return loss on the left side $a_{\mathrm{ret.loss(left)}}$).

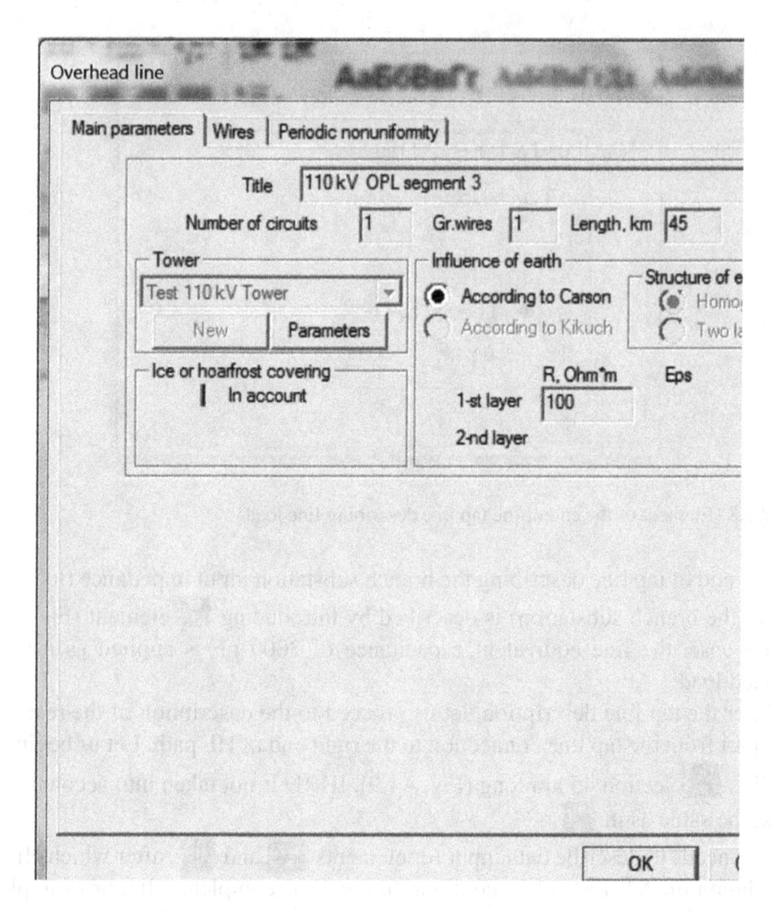

Fig. A1.9 Describing OPL elements after the tap line connection site

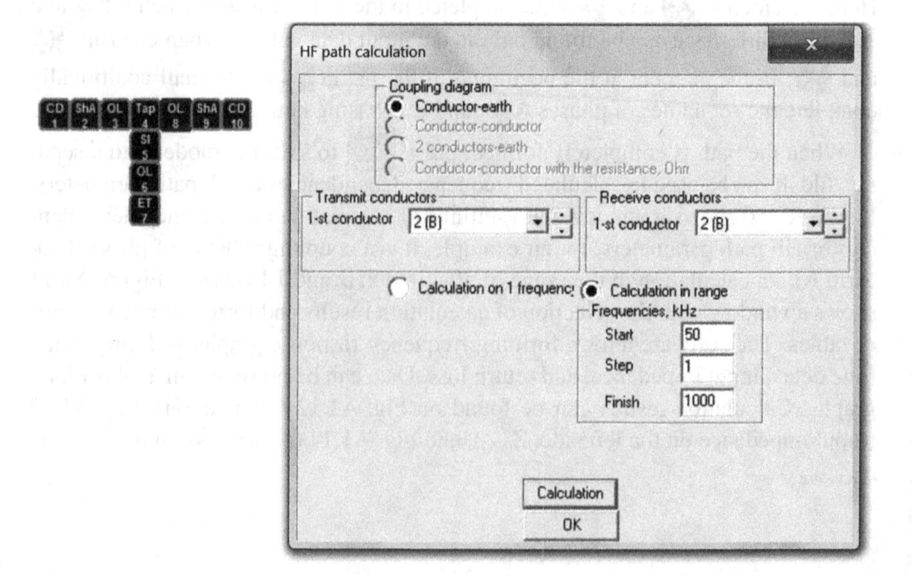

Fig. A1.10 HF path calculation settings window

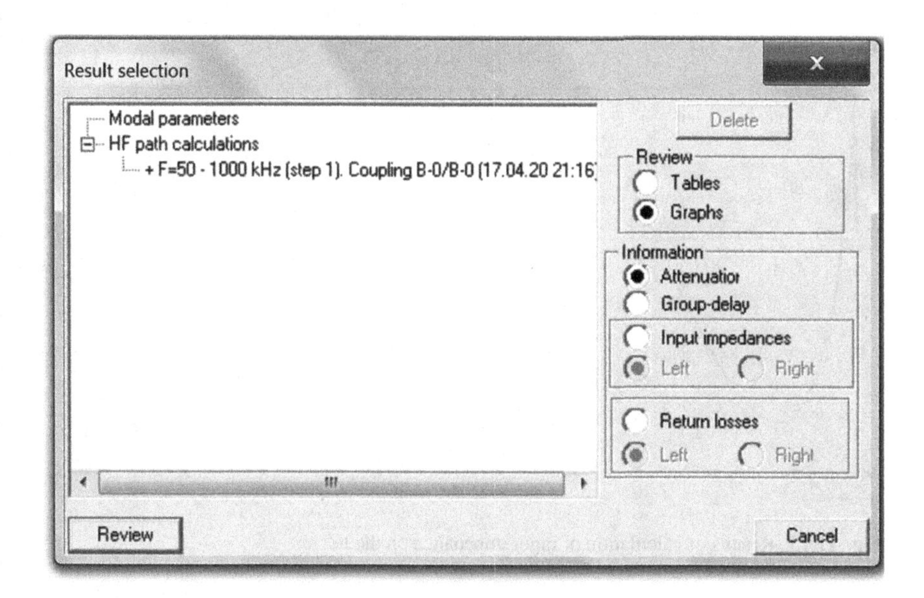

Fig. A1.11 Window with the results of calculation and view of results

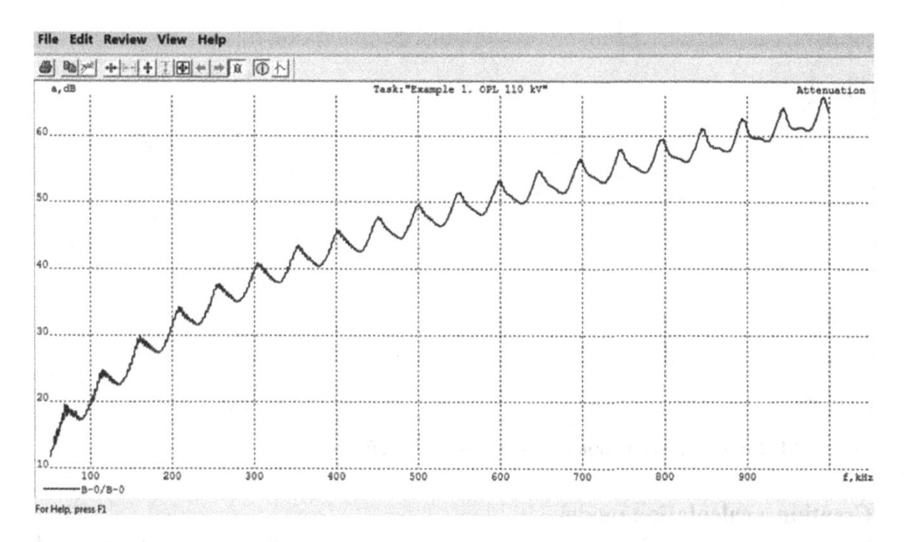

Fig. A1.12 Results of HF path frequency response calculation

Features of creating a calculation model and conduction of calculations in WinNoise

All stages of calculations of corona noise using WinNoise, from the calculation model and to considering the calculation results, generally follow those of WinTrakt. However, there are certain differences. Let us list these differences.

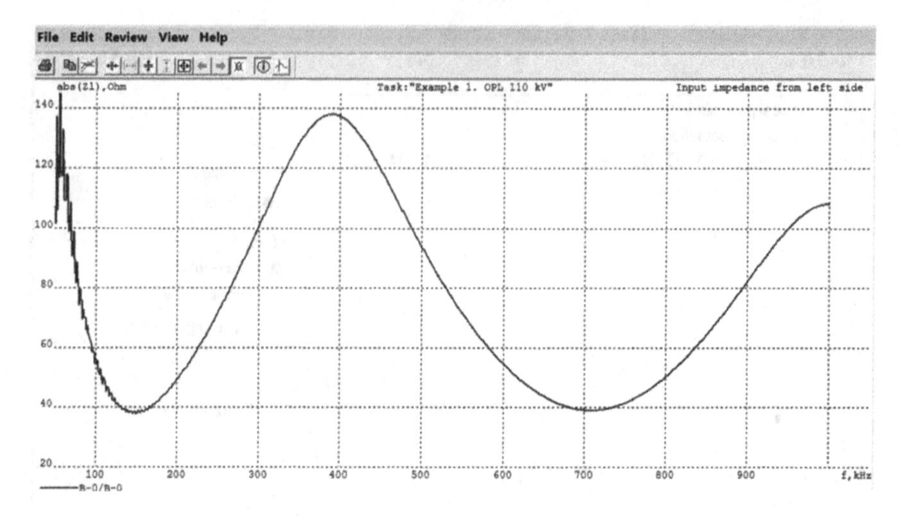

Fig. A1.13 Results of calculation of input impedance on the left

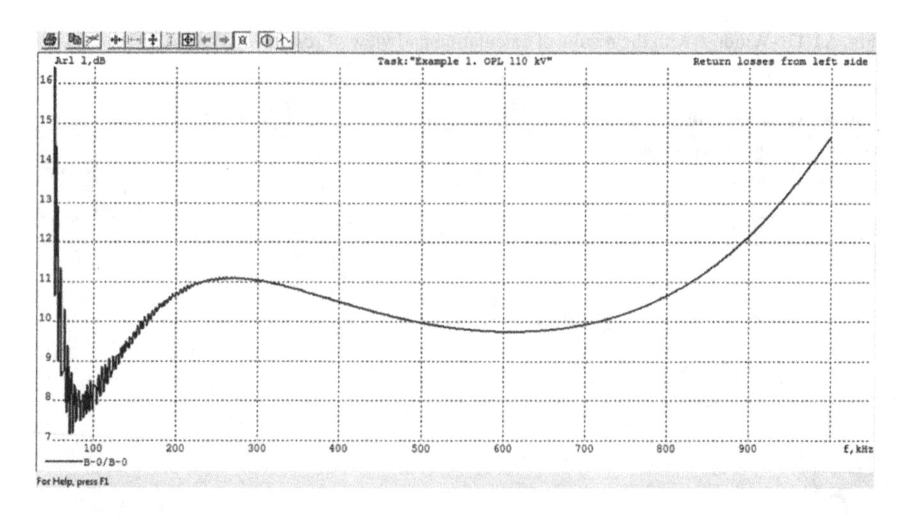

Fig. A1.14 Results of calculation of return loss on the left

Creating a calculation model

Number of elements which can comprise a calculation model is somewhat decreased.

Input data of Total end line load TL and Coupling device CD elements set at the beginning of the path now includes OPLs voltage in kV.

Task for the built model calculation

Calculation condition selection (window presented on Fig. A1.10) has the following differences:

Absence of calculation scheme. The calculation engages all possible schemes of phase conductor coupling with a load of impedance or coupling device.

Method of defining generation function (according to the methodology approved in Russia or CIGRE) is added.

Option of increasing generation function value to a set point on any OPL section is added.

By default, the calculation engages a method approved in Russia, without increasing the generation function.

Calculation results view:

The graphic of calculation results view for selected coupling schemes offers, apart from corona noise level frequency dependences (on both ends of a HF path), a graph of root-mean-square corona noise voltage dependence (compared to average root-mean-square voltage) for the selected coupling scheme and calculation frequency.

Let us consider a HF path (with tap line) calculation model. It is used to calculate the corona noise level and is analogous to the one used above to calculate HF path parameters. This type of model built in WinNoise is the same as the one presented on Fig. A1.1.

Calculation of corona noise on a 220 kV OPL and above is of the greatest interest as it effects a significantly larger impact on DPLC channels than can be expected at a voltage of 35 and 110 kV. Thus, in the provided example, 110-kV OPL with triangle conductor configuration was replaced with the OPL of the same configuration but with the voltage of 220 kV. In addition, phase conductors and coupling devices were replaced with the ones applied for 220 kV OPL (coupling capacitor of 3200 pF; CD bandwidth 78–1000 kHz).

Figure A1.15 shows a frequency dependence of corona noise level (1 kHz band) in a HF path with phase B (upper)-to-earth coupling scheme. The figure clearly shows the way the tap line impacts the corona noise level.

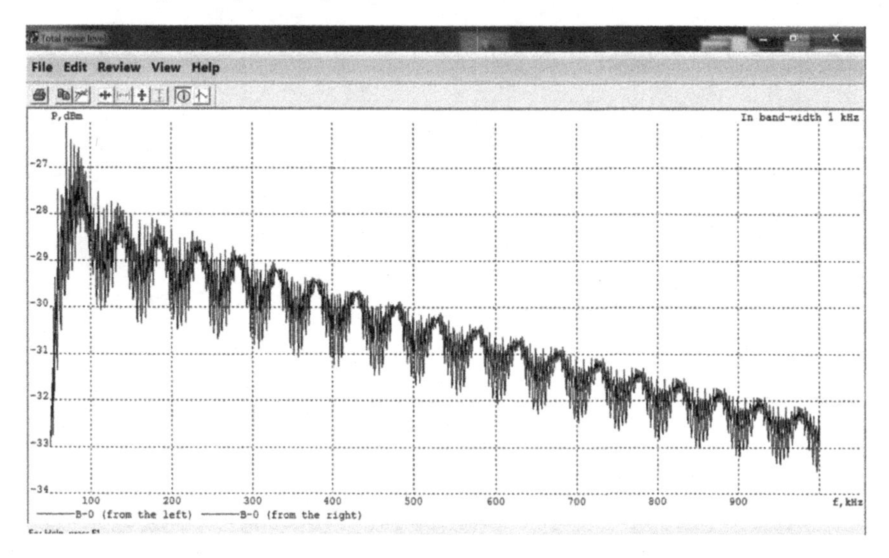

Fig. A1.15 Corona noise levels in a HF path along 220 kV OPL with a tap line

Index

A
Access functions, 9
Additive white Gaussian noise (AWGN), 14
Aluminium conductor steel reinforced (ACSR), 100
Analog signal, 12
Application service data unit (ASDU), 150
Automatic gain control (AGC), 5

B
Bandwidth, 36, 40
Bit rate
 building DPLC system modem, 16
 delay and noise immunity, 16
 dynamic bit rate adaptation, 34–35
 HF modems, 33
 information transfer by DPLC modems, 33
 QAM/TCM modems, 38
 Shannon's limit, 14
 transmission, 41

C
Cable-overhead power lines (COPL)
 attenuation constant, 111
 cable junction, 117, 119
 coaxial, 111
 complexity, 116
 cross-linked polyethylene (XLPE), 111
 design, 113
 effect, 116
 equivalent capacity, 119
 features, 113
 frequency response, 111, 112, 114, 115, 119
 HF bridge, 119
 intra-cable mode, 111–113
 line path, 115, 116, 119
 modal theory, 111
 mode parameters, 112
 OPL connection, 118
 parameters, 111, 116
 phase constant, 111
 propagation velocity, 111
 return loss and input impedance
  attenuator, 127, 128
  frequency dependence, 121, 122
  impedance nominal values, 120
  line transformer, 124
  modulus and phase angle, 120
  operation modes, 123
  parameters, 127
  phase angle, 124
  PLC equipment transmitter, 120
  PLC transmitter, 124
  transmission level, 125
  transmission power, 125
  transmitter, 120, 124, 126, 128
 sheath transposition, 113–115
 single-mode coupling schemes, 114
 wave propagation velocity, 120
Cables, 83
Central processor unit (CPU) module, 7
Channel bandwidth, 15, 16, 38
Channel reliability (ChR), 171
Communication line, 13

A. G. Merkulov et al., *High Voltage Digital Power Line Carrier Channels*,
https://doi.org/10.1007/978-3-030-58365-1

Component, PLC equipment, 7
CPU module, 7
multiplexer module, 8
power supply module, CPU section, 8, 9
voice interfaces, 8
Control, 9
Corona noise, 172–174, 201, 202, 211, 213
features, 130
industrial frequency, 129
integral distribution function, 132, 133, 135
intra-phase coupling, 129
nominal corona noise level, 130, 131
phase angle value, 129
phase-to-earth coupling schemes, 133, 135, 138
phase-to-phase coupling scheme, 136, 138–140
positive, 129
root-mean-square voltage, 130
signal-to-noise ratio, 129
waveforms, 129
wideband noise, 129
Coupling device elements, 205

D

Demodulator performs reverse conversion, 13
Demultiplexer, 11
Digital data interfaces, 7
Digital modulation, 17
Digital power line carrier (DPLC), 145
capital costs, 2
equipment (*see* DPLC equipment)
in modern electric power industry, 1
Digital transmission system, 11, 17, 26
Direct current transmission (DCT), 202
Discrete multi-tone modulation (DMT)
discrete multi-tone, 24
DMT/OFDM modems, 37
modulation, 17
OFDM modems, 24
wideband modems, 39, 40
Dispatch centers (DC), 148
Dispatcher voice channels (DispVoice), 146
DPLC channel
bit rate adaptation
long-term monitoring, 196, 198, 199
PLC channel maximum allowable frequency, 183, 184, 186–195, 197
purposes, 181–183
reliability
EMC, APLC/DPLC channel operation, 179, 180
HF path parameters, 177–179
noise immunity, 171
overlapped attenuation, 172, 173
parameters, 172
required attenuation margin, 174–176
stages, 171
types, 154
packet traffic transmission, 158, 160, 161, 163–165
time division multiplexing, 155, 156
weather conditions, 156
DPLC communication channels, 65
DPLC equipment
ADPLC equipment, 5
APLC unit, 5
as a classic digital transmission system, 11
channel encoder and decoder, 11
communication channel, 15
communication line, 13
as digital transmission system, 12
DPLC systems, 5
error correcting codes, 12
modems (*see* Modems)
multiplexers (*see* Multiplexers)
network element (*see* Network element)
nominal bandwidth, 6
nominal frequency band, 6
PLC equipment pilot signals, 5
structure, 4
system operation, 5
VF modem interfaces, 4
voice and data transfer interfaces, 16
DPLC equipment latency, 36, 38
DPLC Information Traffic
data, 147
features, 146
information transmittes via DPLC, characteristics, 146
singnals, telecontrol system, 147
telemetering and teleindication signals, 147
TM and TI signals, characteristics, 147
types, 147

E

Electric grids, 66
ASDU, 150
data, 148
information frame, 151
information signals, 148
MSS values, 152
MTU, 151–153
PABX, 148
SS and DC, 148

TCP protocol, 152, 154
telecontrol system, 150
VoIP technology, 149, 150
Electromagnetic compatibility (EMC), 179
Error correcting codes, 12, 16, 26, 28, 33, 34
Error correcting encoding, 11

G
Generation function, 202
Groundwires, 83

H
Header compression
compression techniques, 60
cRTP technique, 58
Ethernet header, 55, 59
on DELTA encoding, 55
RFC 1144 VJHC, 56
RFC 2507 IPHC, 56
RFC 2508 cRTP, 56
RFC 3545 ECRTP, 56
in DPLC equipment, 59
IP networks, 54
low bit rate channels, 55
neighbouring packets, 55
NO-DELTA mode in ECRTP technique, 58
on WB-LSB encoding, 56
compression techniques, 56
RFC 3095 ROHC, 57
RFC 4996 ROHC-TCP, 57
RFC 5225 ROHCv2, 57
ROHC v2, 57
robustness, 57
sender and recipient exchange, 55
speech intelligibility, 58
VoIP packet transmission, 58
HF path
characteristics
disturbances, 128
narrowband interference, 141, 142
transient disturbances, 141
Wideband noise, 128
communication line, 65
double-circuit overhead line circuits, 98–104, 107–109
electric grids, 66
frequency response
cable power transmission lines, 78
characteristics, 72
column matrix, 73
coupling schemes, 76
formulas, 76
line path, 72
modal analysis methods, 72
modal components, 74
multimode schemes, 77
OPL transposition scheme, 78
optimal, 78
phase voltages, 75
phase-to-earth coupling schemes, 73
propagation velocities, 77
quasi-single-mode, 78
signal propagation, 73, 77
single-mode coupling scheme, 77
transmitter, 77
transposition scheme, 78, 79
transposition step lengths, 79
voltage, 74
voltage difference, 77
HF bridge, 94–97
modal parameters, 70
attenuation constant, 68, 70
converted line model, 71
converted model, 71
currents, 70
diagonal complex matrix, 69
diagonal matrix, 69
earth mode, 68
framework, 68
frequency characteristics, 69
horizontal phase arrangement, 69
interphase modes, 68, 69
phase current vectors, 68
phase voltage vectors, 68
principle, 71
propagation constant, 68, 69
values of elements, 68
voltages, 70
parameters, 65
PLC channel, 65, 66
power lines, 67
reflected waves, 79–83
special equipment, 66
switching of power lines, 92, 93
TE, 65
terminal equipment, 65
wave propagation, 67
weather conditions, 83–86, 88–92
WinTrakt program, 67
wired communications, 66
HF path calculation model, 202, 209
HF path calculation settings window, 210
HF path frequency response calculation, 211
HF path model formation, 204
HF signal, 4, 5, 7–9

I

Ice hoarfrost and rime deposits (IHRD), 83, 85, 86
Information signal routing, 50
Interphase modes, 84
Inter-symbol interference (ISI), 14
IP networks, 54
IP packet headers, 55, 61
IP Payload Compression Protocol (IPComp), 61

L

Lempel–Ziv–Stac compression algorithm (LZS algorithm), 61
Low bit rate speech codecs, 47

M

Maximum segment size (MSS), 151
Modem latency, 35
Modems
  building DPLC system modem, 16
  digital transmission systems, 13
  DPLC equipment, 16
  DPLC modems bit rate, 33
  modulation
    analog HF signal, 19
    in DPLC technologies, 21
    noise immunity, 19
    phase-shift keying, 17
    QAM, 17, 20
    single-carrier schemes, 17
    subbands, 24
  operation frequency band, 13
  parameters, 16
  PLC equipment, 31
  and power amplifier, 31
  special training" symbol sequence, 32
  wideband, 39, 40
  wired, 30
Multimode coupling schemes, 84
Multiplexer module, 8
Multiplexers
  application in DPLC equipment, 45–47
  basic functions, 45
  data bit rate, 46
  demultiplexing digital signal, 45
  network elements, 46
  prioritization, 45, 48, 49
  processing, speech signals, 47, 48
  signals/network element, 45
  transit of information signals, 49, 50

N

Narrowband interference, 141, 142
Network element
  block diagram, 51
  DPLC channels, 52
  DPLC systems, 51
  Ethernet/IP networks, 50
  header compression (*see* Header compression)
  IP DPLC channel with Ethernet bridge, 52
  LAN ports, 51
  LLQ discipline, 53
  LLQ queue processing discipline, 54
  modern DPLC systems, 51
  OSI model, 53
  OSI protocols, 51
  packet filtering, 53
  packet prioritization, 53
  packet processing, 51
  payload compression, data packets, 61
  PQ discipline, 53, 54
  PQ queue processing discipline, 54
  SNMP monitoring, 62
  Stac compression, 61
  transmitting voice signals, 51
Network management system (NMS), 62
Network protocols, 52
Noise generation function calculation method, 202
Nominal transmission bandwidth, 38

O

Open Systems Interconnection Basis Reference Model (OSI) protocols, 51
OPL elements conductors, 206
Optical groundwires (OPGW), 178
Orthogonal frequency division multiplexing (OFDM)
  DAC unit, 23
  demodulator, 23
  disadvantage, 24
  DMT/OFDM modems, 37
  DMT/OFDM systems, 38
  FDM, 21
  modulation, 17

modulator, 23
modulator and demodulator, 22
multi-carrier modulated OFDM signal, 37
parallel signal, 23
Overhead electric power transmission lines (OPL), 65
Overhead line, 67, 72, 77, 78, 81, 85–87, 89, 91–93

P
Packet filtering, 53
Packet header compression, 52
Packet prioritization, 53
Parameters, 201
Phase conductors, 83, 84
Phase-shift keying (PSK), 17–19, 21
Power line carrier (PLC), 16, 65
communication channels, 1
communications, 1
DPLC equipment, 1
high-quality design, 3
regulation, 3
TDM, 1
technologies, 1
Power supply module for CPU section, 8
Prioritization, 45, 52, 53
information signals, 48, 49
Private automatic branch exchange (PABX), 148
Programming, 9

Q
QAM demodulator diagram, 20
QAM modulation, 20
Quadrature amplitude modulation (QAM), 17
BER *vs.* SNR, 30
block diagram, 19
demodulator, 21
modulation parameters, 37
modulation signal, 18
modulator, 20
M-QAM modulation, 25, 36
OFDM modulation, 17
principles, 17
QAM/TCM modems, 35, 38
signal constellations quadrature diagrams, 19
spectral efficiency, M-QAM modulation, 20
Quasi-single-mode, 103

R
Receiver filter module (RF), 9
Receiving HF signal, 9
Remote monitoring, 51
RFC-TCP header compression technique, 57
Root-mean-square (RMS) value, 129

S
Series impedance, 206, 207
Shannon's limit, 14, 15, 33
Shannon–Hartley theorem, 14
Shunt admittance, 205
Signal propagation theory, 1
Simple Network Management Protocol (SNMP)
implementation, 62
manager and SNMP agent, 62
network management system server, 62
SNMP agent, 62
SNMPv3, 62
Software algorithms, 202
Software implementation, 9
Spectral efficiency, 15, 19, 20, 33, 38, 40, 41
Stac compression, 61
Standards, 2, 3
STO 56947007-33.060.40.045-2010, 3
STO 56947007-33.060.40.052-2010, 3
STO 56947007-33.060.40.108-2011, 3
STO 56947007-33.060.40.125-2012, 3
STO 56947007-33.060.40.178-2014, 4
Substations (SS), 148

T
Tap line, 65, 66, 79, 93, 98, 110
Technological communication, 4
Technological voice channels (TechVoice), 146
Telemetering signals, 147
Temperature, 84
Terminal equipment (TE), 65
Time-division multiplexing (TDM), 1
Transient disturbances, 141
Transit of information signals, 49, 50
Transit stations
delay and speech transmission criteria, 166, 167, 169, 170
simplest model, 164–166
Transmission bandwidth, 38
Transposition, 65, 78, 79, 113, 116, 119

U
User-based security model (USM), 62

V
View-based access control model (VACM) view, 63
Voice interfaces, 8
Voice signals, 50

W
Weather conditions, 83–86, 88–92
Window-based least significant bit encoding (WB-LSB) encoding, 56
WinNoise programs, 201, 202, 211, 213
WinTrakt programs, 201, 202, 204, 207, 211
WinTrakt software, 86

Printed in the United States
by Baker & Taylor Publisher Services